The CRISIS of PROGRESS

John C. Caiazza

The CRISIS of PROGRESS

Science, Society, and Values

LONDON AND NEW YORK

First published in paperback 2024

First published 2016 by Transaction Publishers

Published 2017 by Routledge
4 Park Square, Milton Park, Abingdon, Oxon OX14 4RN

and by Routledge
605 Third Avenue, New York, NY 10158

Routledge is an imprint of the Taylor & Francis Group, an informa business

Publisher's Note
The publisher has gone to great lengths to ensure the quality of this reprint but points out that some imperfections in the original copies may be apparent.

Library of Congress Catalog Number: 2015010463

Library of Congress Cataloging-in-Publication Data

Names: Caiazza, John.
Title: The crisis of progress : science, society, and values / by John C. Caiazza.
Description: New Brunswick : Transaction Publishers, 2016. | Includes bibliographical references and index.
Identifiers: LCCN 2015010463| ISBN 9781412862530 (alk. paper) | ISBN 9781412862073 (alk. paper)
Subjects: LCSH: Progress. | Science and civilization. | Values.
Classification: LCC HM891 .C35 2016 | DDC 303.48/3--dc23 LC record available at http://lccn.loc.gov/2015010463

ISBN: 978-1-4128-6253-0 (hbk)
ISBN: 978-1-03-292912-5 (pbk)
ISBN: 978-1-315-13154-2 (ebk)

DOI: 10.4324/9781315131542

Dedicated to Jo, now and as always.

Contents

There is one great and grave fault in the thinking of American conservatives as well as American liberals. This is their belief in (linear) progress. . . . Liberals adulate Science; conservatives adulate technology. No great difference there. . . . And yet for the first time more and more Americans, some of them perhaps not quite consciously but more and more so, have begun to question the myth of endless mechanical and beneficial "Progress." They are still not an organized or a political minority, but they are not insignificant.

—John Lukacs, *History and the Human Condition*

The rational power needs help in considering, choosing and following the truth, counsel in its election, and knowledge in its fulfillment. It is through this very gift of knowledge, indeed, that we are able to live righteously *in the midst of a depraved and perverse generation.*

—St. Bonaventure, *The Breviloquium*

Pascal's *Pensées* were written about 1660. Many of them are modern not merely in thought, but in expression and force; they would be of overwhelming importance if they were now published for the first time. Such a genius must invalidate the usual conception of human progress. Particularly modern are his rapidity, detachment and intellectual impatience.

—Cyril Connolly, *The Unquiet Grave*

1

Introduction: History and Impact of the Idea of Progress

On Progress, Its Variations, and Rejection

This is a book about the concept of progress, its separate varieties, its current rejection, and how it may be reconsidered—on a philosophical as much as a scientific basis. It not only distinguishes between the concepts of social progress and of scientific progress, but assumes their interconnection. By now, this linkage is well-understood through the medium of the culture: scientists work not only in laboratories, in seminar rooms, and in the field, but in the general ambience of their times, which include the spheres of literature, art, popular music, and philosophy. In dealing with the subject of progress in this book, the main emphasis will be on how science is understood as a result of its direct impact on social values, a view that has been expressed by prominent philosophers. Further, while many historical examples of both scientific inquiry and social and cultural themes will be alluded to or described, the treatment of the subject of progress will be mainly conceptual but also, in a sense, historical.

The four varieties of scientific progress that are described in the following chapters are not meant primarily as a historical description. They are not intended to mean that "reductive progress" followed "enlightenment progress," which, in turn, followed "Whig progress." Rather, the first three varieties arrived at approximately the same time in history as part and parcel of the Enlightenment in the seventeenth and eighteenth centuries, and they continue to have an influence on our understanding of modern science. However, depending on the effect of "historicism," the fourth is the last of the varieties chronologically as it describes the general cultural opposition to the concept of progress at the current time.

Further, it is argued in the following chapters that the current cultural rejection of progress is having a major impact on social ideals. It is no accident that the decline in the concept of progress occurs at the same time that a kind of political quietism is now the condition of Western life; libertarianism is, in effect, the rejection of ideals of progress—either socialist or capitalist—that have hitherto energized state policies for more than one hundred years.

To illustrate that a crisis is at hand, chapter six attempts a description of "where we are now," that is, the state of contemporary Western society in light of the decline of the belief in progress. This is accomplished not through a general tour of politics, economics, culture, and the like, but by examining the effects of technology on Western life. (The descriptions are of American phenomena but apply equally across borders to all Western nations and to the West in general.) It is intended that chapter six leaves the reader with an appreciation that the West and the United States are in a crisis, albeit a slow-moving one. The two chapters that follow constitute a rescue attempt, a philosophic reevaluation of what constitutes progress.

The seventh chapter attempts a reconfiguration of scientific progress based on a criticism of the reductive variety. In reworking the reductive account of the history of science, namely, that it has progressed roughly through three stages—from the physical sciences to the biological and then to the social sciences—it is argued that the scientific method has been increasingly less successful. That is, the mechanical, mathematical, and anti-teleological methodology of modern empirical science is, in general, more successful in atomic physics than it is in sociology, for example. That the social sciences are less successful than the "hard sciences" is not an original point, but the conclusion drawn in this instance is twofold. First, historically, the less modern science can say about a field of research, the more opportunity there is for a recourse to philosophy. Second, as science has proceeded from physics, through biology, and on to the social sciences, it has become less successful in providing explanations.

Chapters seven and eight attempt a partial restoration of the concept of progress in two stages: chapter seven presents the idea that science itself may be seen to progress in a qualified manner, and chapter eight describes what the consequences of such a qualification are in terms of social ideals. Most recently, the arrival of cosmology and of biological descriptions of human nature have impelled the contemporary understanding of social ethics in a manner reminiscent of ancient accounts. Exploring the writings of two of the most influential social philosophers of the present day, Nussbaum and MacIntyre, reveals that both reject the social sciences as a means of providing an adequate basis for a theory of social ethics, and instead admire the position put forth in Aristotle's *Ethics*.

The Entanglement of Science and Social Values

It is one of the major points of this book that the development of modern science is tightly interwoven with modern concepts of social ethics; Galileo's mathematization of local movement was a direct influence on Hobbes, just as the overturning of Newtonian mechanics impelled Popper's philosophy of science and his defense of the open society. The primary account of progress is laid out in chapter two in terms of Whig history, a useful term coined by the historian Butterfield. Whig history is extended, however, from history in general to show the actual progression of modern science, and then to political philosophy as reflected in the thought of Bacon and Dewey. Chapters three and four, which describe "varieties" of progress, riff on the primary historical account laid out in the second chapter, of the Enlightenment view and the Reductive view; the social aspects of these two varieties, as described by Kant and Hobbes, respectively are also presented. The fifth chapter presents various examples of scientific progress, which, in effect, depart from Enlightenment history. "Historicism" denies that science makes progress, extending historical relativism to epistemological relativism; the social consequences are described in the concept of the "open society."

The foremost example of this effect of a newly emerging field of science having a decisive influence on social ethics is Darwin's theory of evolution. Indeed, the influence of evolutionary theory is hard to fully encompass given its fecundity and variety. In Marx's theory of inevitable class warfare, which would lead to a bloody revolution through which a new economic utopia would evolve, Darwinian evolution is thought to justify political theories by both the left and the right. At the same time, Spencer's Social Darwinism theory justifies unrestrained capitalism, in which the poor deserve their fate. These were only the early political effects; later on evolution would be the basis of the eugenics movement, Prussian militarism, and progressive Christian theology. The latest effect is seen in the development of evolutionary theory to explain human behavior in the form of Wilson's theory of sociobiology, which has been attacked on the left as a rationalization of powerful social interests.[1] Lately, however, evolutionary theory is having a depressive effect on progressive politics, since it sets a limit on how much human nature and society can be changed in a permanent way. Through the use of genetic science combined with evolution, sociobiology explains in detail the material basis of human nature, which seems unchangeable and universal. The implied conservatism of evolutionary theory immediately contradicts the progressive mind-set, a fact made even more emphatic by the fact that evolutionary theory has become the iconic representation—even more than relativity theory—of modern science. But on the political right, the reliance of evolutionary theory on traditional values is not enhanced by Darwinian evolution because of its materialistic and reductive implications, meaning that, in effect, evolutionary theory is a dilemma for both sides of the great political divide.[2]

Three Themes

Three themes recur throughout the text, appearing in different chapters and at several points. First, in his Introduction to Stephen Hawking's *A Brief History of Time*, Carl Sagan wrote, "This is also a book about God . . . or perhaps about the absence of God. The word God fills these pages."[3] This is not a unique occurrence; questions about God seem to be an inevitable consequence of dealing with these issues at the current time. This point was made originally by Einstein, who said that he was trying to understand "the mind of God,"—a phrase that is used not so much among theologians as it is by mathematical physicists. While the word "God" does not fill the following pages, it does crop up from time to time as a matter of intellectual necessity; hopefully, the reader will not think that the intent of the book is a disguised form of religious apologetics. (A note of explanation about usage: it has become popular recently to refer to the word "God" in lowercase so as not to allow any preference to one notion of God over another, and, therefore, any one religious tradition over another. However, while lowercasing the noun may be proper usage when discussing gods or religion in general, when the term "God" is used in these discussions it is almost always in reference to the biblical version as understood in the monotheistic religions. Thus, "God" should be capitalized because it is, in fact, a proper name, like Zeus, Venus, and Thor.)

The second theme of this book is that the concept of progress implies an account of the history of science that reflects the work of historian-philosophers of science, including Kuhn, Lakatos, Polanyi, Hanson, and Popper. What can be termed the "new philosophy" of the history of science rests on an understanding of powerful theories as historical presences, not solely as a series of propositions that are roughly consistent with each other. Instead, the term treats such theories as wholes in and of themselves, combining theory with practice, mathematics, metaphysics, ethics, and expectations of future research. This mode of the philosophy of the history of science has been expressed in terms of *paradigms* and *research programs*—concepts that, it is important to realize, concentrate not so much on the totality of science as a historical presence but on its separate fields.

A further aspect of the new philosophy is that when scientific theories are understood in broader terms in which social, cultural, and metaphysical implications are included, they have a penumbra about them that reflects other aspects of culture as well. Newton's system was not only a set of interconnected laws and theories; it also implied a Baconian version of modern science, of large generalizations based on a sufficient number of empirical observations so as to justify or impel an inductive inference. Thus, the universal law of gravitation devised by Newton was not explained by him as a remarkable creative insight but rather as the strict result of the evidence of planetary attractions, as between the sun, the Earth, and the other planets.[4] Further, the cultural impact of Newton's accomplishment was reflected in the salons of France and in the political theory of Locke (a good friend of Newton's), and as the inspiration for Jefferson's belief that the creator God had given to the whole human race a set of inalienable rights.

Finally, the full entanglement of modern science with the general culture of our times is not new and, in fact, is a persistent pattern, as a brief review plainly indicates. For, since the seventeenth century, *new and successful scientific fields have continually provoked new versions of the same thing, namely, a materialistic, anti-teleological, and reductive philosophy that implies a mechanical and deterministic view of human behavior and of social organization.* The most general characteristic of this philosophy is that it is always anti-teleological, denying the validity of any evidence of purpose and design in nature—in effect, to use Toulmin's terminology, denying the validity of natural theology (see chapter seven). The theme originates in ancient atomism, which was explicitly advocated in contrast to Aristotle's teleological account of nature that included animal biology, the human soul, and standards of ethics. Instead, the interplay of unobservable atoms was used to reduce the appearance of any pattern of design in nature. This philosophical materialism denied the reality of lived experience, including colors and sounds, as positive indicators of reality, as well as of free will as the explanation of moral agency.

Ancient atomism established a historical pattern; at the time of the declared departure of modern science from philosophy in the seventeenth century, the ancient materialism was reset into its modern scientific version. Hobbes developed Galilean mechanics into a consistent philosophy of man and of the state, but his version of scientific materialism was only the first. Currently, E.O. Wilson and Weinberg promote their own versions of a scientific philosophy based on sociobiology and on particle physics, respectively. Within one man's lifetime, there are further well-known examples: Monod, whose philosophy was based on biology; Skinner, on behavioral psychology; Freud, on psychopathology; Hawking, on astrophysics; Dawkins, on genetics; and Churchland, on brain science. In each case, these scientists exaggerate the importance of their chosen fields, each of which admittedly offered new and provocative scientific insights into what are recognizably versions of the same theme of scientific materialism. But whether contemporary science is presented to the culture in its evolutionary, genetic, relativistic, quantum, or neurological versions, the means by which modern science is understood to justify itself as the ultimate explanation of human reality and nature as a whole must be judged by standards that exist independently of modern science, that is, by philosophy and by history.

Notes

1. Ullica Sagerstrale, *Defenders of the Truth: The Battle for Science in the Sociobiology Debate and Beyond* (New York: Oxford University Press, 2000).
2. John Caiazza, "Political Dilemmas of Social Biology," *Political Science Reviewer*, Washington, Vol. XXXIV, 219–59.
3. Stephen Hawking, *A Brief History of Time* (New York: Bantam, 1988), x.
4. Isaac Newton, *The Principia: Mathematical Principles of Natural Philosophy*, trans. I.B. Cohen et al., (Berkeley, CA: University of California Press, 1999).

2

Whig History and the Progressive Society

Whig History and Critical History

Whig history is generally understood as a portrayal of past events that leads inexorably to a future state that is identified with the present, or, more likely, a projected end point that is a readily perceived extension of the present. The end point, whether realized in the present or in a future projection, is understood to be a desirable end, a maximization of human and historical potential. Therefore, the process of historical development is seen as a good thing. The term "Whig history" was used initially to describe the history of English governance in its parliamentary and other aspects, as a progression from monarchical tyranny to the self-governance of free men and women.[1] It was significant in the political sense that Whig history predominated at the time when, during Queen Victoria's reign, the British Empire was being extended to its farthest limits geographically and culturally—from India to Kipling's poetry. The sense of historical inevitability allows the philosophically inclined observer to make value judgments about events and personages; in general, Whig history is optimistic not only about the trend in history, but also about the general ability of human beings to control their own destiny. As a result, it has the unfortunate tendency to identify historical actors as either good or bad depending on whether their actions advanced or impeded the path of progress.

Applied to the history of science, a Whig interpretation tends to arise unreflectively; for example, Newton's comprehensive theory is seen to succeed that of Kepler, Descartes, and Galileo. Just as British historians of the nineteenth century could look with satisfaction at the parliamentary and imperial success so evident in England at that time, so might they also look with satisfaction at the succession of great English accomplishments in science, with such names as Dalton, Newton, and Darwin. And, in science as well as in politics, the Whig view applied value judgments to those people and theories according to whether they pushed forward or impeded the course of scientific progress; phlogiston was condemned as bad science, while evolution was a "triumph." Yet, as we know, the British Empire failed after World War II. And a larger view of history reveals that even the greatest of empires, such as Rome's, fail, as English public schoolboys well knew as they read Caesar and walked on old Roman roads that are still in use in England today. In retrospect, the British Empire was a onetime thing, a project that the British could not, and, in the end did not, want to sustain. But the same is not true of modern science, for its progress is continually sustained and seemingly has not faltered. Newton may be dead, and his mechanical theory overcome by relativity, but the Englishmen Hawking and Dyson have succeeded him.

The career of modern science has come under the same kind of criticism as did British imperial history. Indeed, the link has been made by critical historians between British hegemony and the physical means by which it was made possible, that is, by repeating rifles, cannons, and explosives. These observations extend to the general sense that the link between modern science and Western political hegemony is not accidental but intrinsic, and thus modern science has become defined as the means by which Western power was acquired and is sustained. Whole schools of science studies, as well as the sociology of science, have arisen to identify its sins of racism, patriarchy, elitism, and Western hegemony, so that it may seem to be no more than the means by which political power is employed.

These "new schools of resentment" may be seen as a reaction or an overreaction to the prior assumption that modern science was the prime mover of Enlightenment progress—not just an indicator, but an essential cause of the rapid success of Western culture from the fifteenth to the twentieth centuries. The era of self-congratulation is now over, replaced by one that denies that the notion of progress is a valid description of Western culture and of modern science. The issue is not only one of historical interpretation, but attaches also to the epistemology of science. The objective of recent scientific studies is not only to assert that science has been the handmaiden of Western domination, but that modern science itself does not have direct access to physical reality, and, therefore, cannot be relied upon to be the truth-teller of advancing world civilization.[2] In effect, contemporary critical approaches to the history of science deny that science makes progress and provides a revisionist/critical reinterpretation of what have been thought to be its bona fide victories and advances. Therefore, these approaches deny the premises of the Whig history of science.

Evidence of Progress in Science 1: Discovery—Spectrography

While Whig history may be discounted these days, it is apparent that there is a difference between the history of empires and the history of science. Despite the claims of its recent critics, modern science does progress, and, in the face of the criticism that asserts otherwise, it is important to make this case. Historical examples provide sufficient means to prove that science is a progressive enterprise in which prior events influence subsequent events, and provide a greater theoretical understanding of specific areas of the physical universe.

Science has two fundamental parts: empirical discoveries that accumulate over time as more facts are discovered by observation and experiment, and theory, which proposes general explanations for the phenomena discovered by experiment. Following this distinction, two accounts will be given of how science progresses in the areas of empirical discovery and of theory. The first example considered will be that of the technical invention of the spectroscope, which has led to advances in optics but also in astronomy and atomic physics; the second will be the laws of genetic inheritance, a hugely important theory about the generational transmission of bodily characteristics. (The distinction between empirical fact and theory is not intended to be absolute; there is constant overlap between the two in the practice of modern science, as will become clear in the following two examples.)

The history of the spectroscope is a revealing instance of how a commonly observed phenomenon, known to ordinary people and reported in ancient times, can be turned by a succession of refined observations and experiments into a major scientific resource. Looking at sunlight through the edge of a piece of broken glass shows that it has more colors in it than white. That and other such common phenomena as light passing through a glass bric-a-brac set on a kitchen table sometimes produces the same effect. Another example is the effect of sunlight on frozen snow, which can produce a myriad of small dots of light in the different colors of the spectrum including red, orange, blue, and green (personal observation). There may have been experiments in ancient times that attempted to reproduce this spray of colors by means of an arrangement of glass or water, but they have not been recorded. Yet, the effect was well-known enough that prisms were being produced to capture it in the sixteenth century as optical science was being developed, first by the Arab philosophers and then by Europeans. Refraction and the angle of incidence through various translucent substances was the subject of experiments at the time. The prisms in use were made of triangular pieces of glass through which sunlight was passed, producing the well-known spectrum of colors—the colors of the rainbow, in fact, in a regular and gradual manner. Newton turned his eye toward the area of optical research before he produced his works on mechanics and gravity. He ran a number of experiments involving light, the most famous of which was using the glass prism in a darkened room to get a clear result of the solar spectra that were plainly demonstrated on a white board. Newton later used a second prism to collect the spectrum and to condense it back into white light. However, this was only a first step.

The next significant step was the discovery that light that was artificially produced by combustion also produced spectrums. Experimenters manufactured devices made up of a chamber for burning a small quantity of matter connected to a viewing tube containing a glass prism; by that experimental means a spectrum unique to that sample would be observed and recorded. With this apparatus in hand, chemical experimenters burned various types of matter at the same time as they were becoming aware of an increasing number of elements and compounds. The spectrograph, as we may now term it, was used along with the laboratory-bench techniques of combustion, condensation, distillation, and other methods to establish the existence of separate chemical compounds and their constituent elements. Sodium, oxygen, hydrogen, carbon, and other elements were discovered between the sixteenth and the eighteenth centuries. What was also discovered was that each separate element had its own individual spectrum that included several dark lines, a fact that was of major assistance in analyzing the makeup of compounds and mixtures by identifying their separate elements.[3]

Another significant application of early spectroscopy was to astronomical observation. The light from the sun and also the stars produced unique spectra from which the actual composition of these "heavenly bodies" could be discovered. (Helium was first discovered as an unidentified spectrum from solar spectroscopy, and then only subsequently discovered to exist on earth.) Both the chemical and astronomical applications of spectroscopic analysis were to lead to important results later on.

As the observational techniques were further refined, Fraunhofer analyzed the lines that appeared in spectrums of, for example, hydrogen and most other substances subject to analysis. The lines that demarcated the point where one color sheared off into another in a spectrum gave indications about the identity and nature of the substances. Fraunhofer also made significant advancements, first, by developing means of making prisms, the glass of which was clear of impurities and precisely ground to measure; and, second, by replacing glass prisms altogether with his invention of diffraction gratings. The diffraction grating resembles a metal file with very fine grooves. First made by wrapping thin wire over a small board, it provided such a degree of distinct separation among the spectral colors that for the first time the length of light waves could be distinguished and measured. The humble Bunsen burner known to high-school chemistry students was a significant improvement in spectroscopy, as it burned with a clear blue flame that did not infect the samples being burned and analyzed. Bunsen and Kirchhoff made refinements in spectrographic techniques that allowed reliable identification of the ratio of elements within complex compounds. By now, spectrography was not only helpful in the discovery of the elemental constituents in complex chemical compounds, but also provided strong indications of the existence of hitherto undiscovered elements.

Spectrography eventually fulfilled its destiny by, aiding in the early twentieth century's two great general discoveries in physics: relativity and quantum mechanics. In astronomy, the study of the spectra of stars and distant galaxies by means of the spectroscope gave evidence of both a "blue shift" and a "red shift," which could reasonably be interpreted to indicate that far-off stars were either traveling toward our galaxy, or away from it. Hubble famously used spectrographic analysis to determine that the red shift became more pronounced the farther away the galaxy under observation appeared; he encapsulated his discovery into a law that provided evidence that the universe was expanding. It was also possible to deduce from Hubble's expansion thesis that the universe had a beginning in time, or a point of origin in space-time—the "big bang" famously explained by Einstein's theory of general relativity.

The discovery of Fraunhofer lines in the nineteenth century was linked in the early twentieth century to experiments relating to energy states, when a sample was excited by heat or other forms of energy applied to it. Einstein, relying on Planck, had discovered that the energy released as a result came in discrete packets of energy. The quantum theory of light emission was then connected by Bohr to the energy levels of an atom as it underwent excitation; this relationship, in turn, was connected to the model of the atom in which the nucleus is surrounded by electron shells. The final triumph of spectrography was its role in the realization that the exact appearance the Fraunhofer lines was the direct result of the sequence in which electrons "jump" instantaneously from one shell level to another, during which a light quanta (or a photon) is emitted. This condensed history of spectrography (which leaves out such development as mass and X-ray spectrography) is useful to show the advancement of science in its empirical mode; from the common observation of colors emanating from sunlight on frozen snow and from broken glass, to the origin of the universe and the internal structure of the atom.

Evidence of Progress in Science 2: Theory—Mendel's Laws

The advancement of spectrography involved not just an accumulation of raw observations but also a lot of theorizing, which was necessary to explain effects and to provide further lines of research. However, another historical example is useful to illustrate the advancement of the theoretical side of science, that is, the discovery of the laws of genetics by Gregor Mendel.

Mendel's laws of inheritance are well-known from their inclusion in high-school biology. It might be said that Mendel's laws are merely laws, and thus lack the comprehension that is essential to theories that operate at a higher level of abstraction. Mendel's discoveries, however, are not simply a set of low-level inductions. His laws of inheritance provide, in general terms, a theory about the inheritance of characteristics that are manifest in the soma of plants, and, by extension, in animals and in any other form of living thing that is beyond the level of single cells and viruses. Namely, this theory posits that (1) somatic characteristics are the result of particularate and identifiable causal entities (genes) that pass from one generation to another; (2) these causal entities possess a kind of two-valued logic that expresses itself either as present or not present in the soma (although the entities may also exist, but not visibly, within the soma); (3) living things are composites of the presence of these causal entities; and (4) this process of character inheritance by particularate entities is the cause of the variation necessary for the operation of evolution, for these variations provide the somatic characteristics upon which natural selection works.

As a theory or a set of complex laws, Mendel's discovery appears picture-perfect, which leads, oddly, to an issue the response to which is difficult since Mendel seems to have no predecessors in the field of genetics. Mendel's discoveries are an example of how science should be done at its maximum efficiency, the conducting of relatively simple experiments over three generations of plants leading to a complex abstract theory. One indication of how well Mendel did his science is that the laws as he presented them remain substantially unaltered; the ratios and the terminology remain just as he originally wrote them down, and are as well-proven as most scientific discoveries ever are. Indeed, one later geneticist suggested that the experiments were too perfectly done, and that Mendel must have "cooked the books" in order to get the evidence for the 3:1 ratio of dominant-to-recessive genes that he discovered. But then it may be asked, where did Mendel come up with the idea of dominance and recession and their 3:1 ratio to begin with? It wasn't as if there were textbooks of genetics that he could plagiarize.

There is a dramatic aspect to Mendel's famous discoveries that always seems to come through in even the driest presentation of the laws of genetics. The basic historical fact is that one of the great discoveries in the history of biology went unrecognized for two generations. As Eiseley noted, "thirty-five years were to flow by and the grass on the discoverer's grave would be green before the world of science comprehended that tremendous moment."[4] Mendel is presented as tragic figure that never got his due recognition; he was almost as great as Darwin in the advancement of biology yet was ignored during his lifetime. The reasons for this have been probed: Was it anti-Catholicism or a general prejudice against religious interference in science, the condescension of experts like Nageli to the monastic amateur, or the obscurity of the local scientific journal in which the discoveries were presented. (Yet, researchers found the papers readily enough when they needed them.) Here, however, we take the on the frequently asked question of why Mendel's discoveries were ignored, not in biographical or historical terms, but as a singular case of how a theoretical advancement can be so far ahead of its time that it is literally incomprehensible to the major practitioners in an established scientific field.

Mendel resembles a modernist artist, an expressionist painter like Van Gogh or Gaugin, who is ignored in his lifetime and later is recognized as not only a great artist but the founder of a new school of artistic expression. Indeed, the comparable issue is that of *expression*—from visual, artistic expression to the expression of genetic forces in the physical characteristics of plants and animals. In addition, like the artistic examples, Mendel's "laws of plant hybridization," as he called them, arise in a dramatic, almost explosive, fashion, but only when their time has come. What was special about the nature of his discoveries that effectively hid their significance from the world of biological science for almost two generations? The answer is that Mendel's discoveries, which, while based on the most patient run of experiments (over seven years) and of controlled observation, were explained by him in terms of an invisible and empirically undetectable cause, an "inferred entity" that eventually was termed the "gene."

Consider the evidence, or lack of evidence, provided by microscopic cytology at the time that Mendel made his discoveries. In the seventeenth century, Van Leeuwenhock invented the microscope, the use of which early on aided in the discovery of the cellular nature of protoplasm; only a slight magnification revealed that human skin is made of up discrete, observable, irregularly shaped but self-contained "packets." Further, in 1839 Schleiden and Swann discovered that all living matter basically occurred in cellular form. At the time when Mendel devised his experiments, microscopy enabled observers to see only the cell as a whole. Note that chromosomes were not yet observable because they could be seen only under high magnification, and then only when staining techniques enabled the observer to clearly discern their shape and existence. As for the genes that constituted the individual chromosomes, they were as yet unknown. What was known was that somehow characteristics were transferred from one generation of plants and animals to another by means of the marriage of sperm and egg. Although these things were themselves very small, yet not microscopically so, they contained all the information needed to develop a new offspring that was only nascent within the sperm and the egg. Darwin supplied the theory that, as it passed through the circulatory system, the blood collected in an unknown fashion the information about the parents' somatic construction, which was then passed reproductively to the offspring. The point is that in order to explain plant hybridization, Mendel had to move through three ontological levels—from the cell, to the chromosome, to the gene. However, of the three, only the first level was known to science.

Mendel's genius was to patiently think through the process of the transmission of observable characteristics through three generations of pea plants, and to infer not merely the unobservable structure that underlies the cell (the chromosome), but the structure that underlies the chromosome (the gene). Thus, Mendel's theoretical inferences were correct and they provided the solution to the mystery that had confounded Darwin. But because he had drilled down through the two levels of cellular reality and theoretical comprehension that were then unknown, Mendel became an unrecognized prophet whose solutions were discovered only after he died.

The point of presenting the two examples of spectroscopy and genetics is to indicate that despite postmodern criticism and skeptical attitudes, science does make progress because it advances in both the empirical and theoretical areas. Thus, Whig history of science, whatever its perceived deficiencies and possible alternatives, will always retain its major plausibility. Whig history of science has a further advantage besides describing science's inherent progressivity—namely, that it delineates both a beginning and a projected end point in the historical career of modern science. The following four sections will deal with the Whig description of the beginning of modern science, of the anticipation of its end point, and of its inherently social nature.

The Separation Thesis: Modern Science Departs from the Medieval World-View

In the Whig version of its history, it seems as if modern science begins in the sixteenth century, when it separates from philosophy, although the absolute start of things is left vague. This may be called the *separation thesis*. Often, the history of modern science is considered to have begun with Galileo, in the late sixteenth and early seventeenth centuries, but it can be pushed farther back, to Copernicus, whose book, *De Revolutionibus Orbium Coelestium*, published in 1512, may be said to have started it all. Then there are the prophets who predicted the rise of this new form of knowledge, particularly Francis Bacon and Thomas Hobbes, whose writings appeared in the seventeenth century. In any case, Whig history implies that there was a definite separation between the history of modern science and the medieval forms of knowledge, roughly meaning the philosophy put forth in the thirteenth century by major theologians such as Thomas Aquinas. Aquinas's majestic intellectual creations were characterized by Voltaire, among others, as based on the fruitless distinctions and crabbed speculations of narrow-minded theologians.

Historians of science in the nineteenth century, which was also the century of Whig history of Anglo-Saxon politics, included Whewell, White, and Draper, who emphasized the notion of a sharp demarcation between the old, medieval form of thinking, and the new scientific form.[5] The chief historical figure in these presentations was Galileo. But the issues surrounding the Galileo case were not merely astronomical, as in the debate between the geocentric and heliocentric alternatives, which Galileo termed "the two chief world systems" in the title of his great expository defense. By calling them "systems," Galileo indicated that much more was at stake than merely which provided the true version of the relationship of the motions of the sun, Earth, moon and other planets; for what Galileo was facing was not an astronomical theory but a worldview. Thus, there was the sense that something of major importance happened beyond an episode of scientific theory change. What Galileo opposed was a complicated and well-integrated system that encompassed astronomy, mechanics, philosophy, practical ethics, and, not least, religion that constituted a comprehensive intellectual, religious, and social whole—a gestalt, or worldview.[6]

In the medieval worldview, all these varying "memes" were interconnected: the human person to the heavens, the four elements to neo-Platonic philosophy, the Bible to Aristotle's *Physics*, the four seasons to agriculture and astrology, the work of stonemasons to the glory of God. Above all, the interconnectedness of the medieval worldview gave the cosmos a homey aspect that was conformable and supportive of a powerful tendency to explain observed phenomena in teleological and religious terms. Of all the causes that have been offered, from the sixteenth century onward, to explain the difference between the medieval and the scientific worldviews, the replacement of teleological explanations with mechanical ones and the deliberate expulsion of "occult" causes are the most relevant.

Ideologically, the medieval world picture constituted three parts. The first was Christian doctrine as interpreted by Catholic and then later by Protestant theology, which were both based on the traditional exposition of the Christian religion, the fifth-century Nicene Creed. This creed describes God as Trinitarian with the Father as the "creator of all things, visible and invisible," and asserts that all of mankind is subject to divine judgment; the Nicene Creed is usually said today at most Christian worship services of all denominations. The second important part of the medieval world picture was Aristotelian philosophy, which medieval thinkers, including Jewish and Muslim philosophers, had attempted from the eleventh through the fourteenth centuries to integrate with their versions of the theology of revealed monotheism. Aristotle's thought represented the ultimate authority of secular knowledge, the pinnacle of what the human mind was capable of knowing, on its own—that is, without the supplemental revelation of the Bible and the *Koran*. The third part was geocentric astronomy, which, when combined with Greek philosophy, gave a picture of the universe as literally centered on the Earth, which, in turn, was surrounded by the spheres corresponding to the four elements of earth, water, air, and fire. Beyond the confines of the Earth there existed a succession of spheres that corresponded to observed "heavenly" bodies, including the moon, the sun, and the seven planets, which were observable to the naked eye. Following the tenets of Aristotelian cosmology, a final sphere of the "fixed" stars corresponded to the heavenly vault presumed by Christian theology. It was a tightly defined universe, almost comically small when compared to our current picture of a universe composed a vast swath of space-time, approximately fourteen billion light-years across.

Interestingly, the foremost exposition of the medieval worldview is not found in a philosophic text or a theological tome, but in Dante's *Divine Comedy*. The epic poem consists of three parts, the most famous of which is the *Inferno*. Dante's description of the physical environment in which he and his guide, Virgil, travel vividly reflects the knowledge of nature of the time: the exactitude of the description of Hell as a deep spiral pit in the center of the Earth is so precise that studies of it were subsequently conducted by mathematicians and students of nature, including Galileo. The poem also reflects, in an indirect way, Homer's *Odyssey*—the most plangent aspect of the great poem is its reflection of the search for the meaning of life by a man who is approaching the crisis of middle age.

Leaving Aristotle, One Science at a Time

The medieval worldview has its partisans and devotees even now: an entire medieval monastery was once disassembled down to the stones and reerected in a museum in New York City, while the theology of Thomas Aquinas remains the semiofficial orthodoxy of the Catholic Church. But in the terms of modern science that we now take for granted and that characterize our worldview, we can pursue the *separation thesis* to point out how the overturning of Aristotle's philosophic dominance led directly to the formation of modern science.

Toulmin points out that the dissolution of Aristotelian philosophy was effected by the growth of specialization among those scholars who studied the natural world.[7] During the middle and late medieval period, there was such an accumulation of empirical data about plants, animals, astronomical phenomena, and the Earth itself that specialization became a practical necessity; no longer could any individual scholar claim all knowledge as his province. (Leonardo da Vinci was perhaps the last man who could make that claim.) But the main cause of the great separation was the realization that Aristotle's science did not provide an overall and uniform explanatory basis for the new sciences that were being developed. The departure from Aristotle's intellectual dominance to the establishment of the modern scientific worldview happened in stages. Specific scientific fields were devised by scholars, and the entire intellectual process took about six centuries—spanning from the fifteenth to the twentieth centuries. The transition has been characterized as a family feud in which Philosophy as father and poetic Myth as mother established a pleasant and enduring intellectual home, despite which, one at a time, the children left to set up households of their own. The fields that have separated off from the ancient intellectual castle keep are as follows:

Physics—or, more exactly, mechanics or the theory of motion—replaced the notion that a body lost its initial force as it moved until it simply stopped moving; the theory of inertia that had developed by the fifteenth century rebutted this aspect of Aristotelian science.

Astronomy departed when the heliocentric thesis was fully developed by Copernicus in his *De Revolutionibus* in the sixteenth century. The cultural effect was a sundering of the iconic representation of the medieval conception of the universe, since the Earth was no longer at the center.

Biology emerged in the nineteenth century when Darwinian evolution rejected a teleological account of natural processes, replacing the special creation of each individual species with a theory of (slow) development caused by natural selection and variation.

Psychology, sociology, and political science, in the late nineteenth and early twentieth centuries, were developed under the single category of "social science" by the invention of a variety of sometimes conflicting theories that included behaviorism, psychiatry, and the quantitative description of the behaviors of populations. The general direction was toward replacing mental deliberation and freedom of will as putative causes of human behavior with empirically and quantitatively describable ones.

Modern logic is often overlooked as a major departure point because its development occurred mostly during the twentieth century, when the transition from the medieval to the modern scientific worldview had already been accomplished. Aristotle's invention of the syllogism, however, had dominated the field of formal logic for over 2,000 years but was eventually replaced by new forms of symbolic and mathematical logic created by such figures as Frege, Boole, and Russell.

All these separate fields departed from the Aristotelian household in which they were born and nurtured in order to set up academic households of their own and to spawn their own sets of offspring. Each field has, in turn, become ferociously independent, with little reverence for or memory of the philosophic and poetic sources that gave them birth. In fact, like ungrateful children, scientists tend to be resentful and scornful when they remember these origins—even as they sometimes return to the old man for intellectual loans. As Einstein once remarked, quantum discovery bears an uncomfortable resemblance to Berkeley's "*esse est percipe*." And how different is the concept of "alternate universes" from Plato's ideal mathematical forms?

Beyond the separation from the Aristotelian worldview through the development of autonomous fields of study, the most general separation in terms of scientific explanation centered on the subject of causality. Aristotle's analysis in his *Physics* states that there are four causes—material, formal, efficient, and final—and gives a detailed, complex account of how each type, all of which are tightly related, must be accounted for in the explanation of any single phenomenon. The details of his analysis are perhaps less notable, however, than the fact that it contains more than one type of cause, while modern science, it is fair to say, assumes there is only one.[8]

The reduction of causality to a single type provides a means of intellectual and explanatory efficiency that is preferred by modern science, but that even so has its own level of complexity. Because scientific causality retains the two-sided character of science in general, namely, that it is both empirical and abstract, it appears to be both mechanical and mathematical in nature. The mechanical aspect is apparent in what Hume criticized as "billiard ball" causality, that is, that causes are a matter of physical contact between bodies. It was on the basis of this clumsy idea of mechanical cause that the Cartesians criticized Newton's theory of gravity. They realized that heavenly bodies appear to move in their assigned orbits by themselves, without a body to push them along; there is no astronomical cue ball in contact with the planet Jupiter. So the Cartesians explained the motion of planets as resulting from the impact of "vortexes," which were material enough to provide a mechanical impetus but which, for whatever reason, were invisible. Newton, on the other hand, explained the orbits of the planets by means of a mathematical formula; when tasked with providing a mechanical cause, he replied "*hypotheses non fingo*," which meant, in effect, that it was enough to have provided a very accurate means of predicting planetary motion without speculation as to its mechanical cause.

It could be argued that gravity has never been successfully explained by a mechanical principle: Einstein's relativistic field equations are obviously more abstractly mathematical than are Newton's laws of planetary motion, which were used, in turn, to explain Kepler's three laws and to predict the orbit of Halley's Comet. Science retains, as it were, the desire—a reminiscent hankering—for explicitly mechanical causality in the sense of one body impacting another; increasingly science relies on mathematical expression. The mathematics more often utilized by modern science from the late nineteenth century onward however was not the deterministic calculus but probability mathematics which was useful for the development of gas laws and other applications, including fluid mechanics. It provides a means of prediction and control of natural phenomena that are inherently chaotic. The development of heat theory and of research into the behavior of gases would require mathematical methods that extended beyond the deterministic calculus to include the use of probability mathematics. The mathematics of probability is also used extensively in the social sciences including epidemiological study of human diseases. But the reliance on probability methods, including sampling, detection of correlations, and regression analysis, further increases the distance between the mechanical and the mathematical aspects of scientific causality.

The Projected Ideal Ends of Science

Whig history of science is based on a vague but insistent belief that modern science had a beginning in time; it also maintains that it has, in one sense, an end that is not only a projected end point when scientific knowledge reaches a certain climax, but also, in a second sense, an end that lies in its purpose of understanding and of controlling the forces of nature for the good of mankind. The first, or what may be called the theoretical, end point of science is that it will be reached in the form of a vast, complex theory, following the example of Newton's *Principia*. So successful was Newton's accomplishment, measured not only as a finished mathematical project but also by the fact that other scientists used its results to extend physical knowledge in a variety of ways and areas, that it remained the climactic representative of scientific knowledge for about two centuries. Its replacement by the theory of relativity and by quantum mechanics was, in fact, traumatic for the Enlightenment view of what constitutes scientific progress and the power of human reason.

Given the increasing complexity of physics, as discoveries of such phenomena as radiation and electromagnetism that were made after the breakthroughs of Newton, it has become apparent that a last, grand theory will necessarily be a *set* or a collection of theories, of various levels of comprehension and abstraction, that will all be subsumed under the rubric of one final theory. But this possibility is limited to the phenomena of physics only; it is not clear or obvious how the phenomena studied by biology and the social sciences would be linked or reduced to those studied by physics. The implied ambition of science is to be able to explain all the phenomena of the physical universe, and to reduce those residual parts of the universe and of human experience that purportedly lie outside the realm of science to a form of physical activity and nothing else. Such is the very large, not to say hubristic, aim of scientific ideology. While such completion is difficult enough in the realm of physics, when the intention of scientific ambition is to encompass all of physical reality, including those areas discovered by biology, the final end of science begins to seem very unlikely if not impossible.[9] Such considerations, however, have not up to this point dimmed the sense that the final end of science will be attained in the form of a set of equations that, according to one contemporary physicist, will be condensed yet comprehensive enough to fit on a T-shirt.

But accomplishment of the "theory of everything" by mathematical physics would also have social aspects that would, in effect, be the final vision—not only for reductive physics but for mankind as a whole. A sense of what such a projected end entails is given by Peirce, whose speculative account of the ultimate end of science is "asymptotic." That is, that as scientific knowledge proceeds, it gets closer and closer to a description of ultimate reality, but never fully attains it. Peirce's idea of a possible if not likely epistemological climax influenced Teilhard de Chardin, whose expression of the culmination of knowledge was not only scientific but social ("noosphere"). This culmination was also mystical, and was described as the "Omega Point," that is, the Second Coming. While determinedly anti-materialistic, Teilhard de Chardin's mysticism is an indication of the general idea of culmination entailed by the concept of "the end of science."[10]

Since the intended end of science is so comprehensive, it is also, by implication, a social enterprise. It may have been thought that the Whig version of science history could be portrayed effectively in the manner of the "great man" theory of political history, concentrating on those figures who accomplished the great events and transformations such as Napoleon, Caesar, and Lincoln. In like manner, Whig history has sometimes been portrayed in the accomplishments of single characters, such as Galileo, Newton, and Darwin, whose accomplishments stand out. But once an observer looks at it in greater detail, it becomes apparent that the great man account is deficient in that transformative events took place in an emphatically social context apart from which the careers of such great men cannot be understood. Thus, despite the glorification of Galileo's career, which is commonplace in the Whig history of science, the great man lived in a social context as an ambitious university professor, celebrity scientist, member of the Academy of the Lynx, and an inventor of military technical apparatuses. Further, as a Roman Catholic, Galileo consorted with high-level ecclesiastical officials. Because of the Whig emphasis on Galileo as a great man, the discovery in the late nineteenth and early twentieth centuries that he had predecessors in the mathematics of motion came as an unwanted surprise (see the following chapter). This is also true in other cases. For example, historical observers, including scientists themselves, want to know who influenced Einstein, since the discoveries of the two theories of relativity seem to have come from him alone, in isolation from what was being discussed among physicists at the time. The contrast between relativity and quantum mechanics, in particular, provokes interest in this question, since the history of the development of the latter is so obviously a social event that it cannot be limited to one name; Planck, Bohr, Heisenberg, Schrödinger, de Broglie, Feynman, and Einstein himself all played essential roles.

Bacon's Prophetic Vision in the *New Atlantis*

The inherently and necessarily social character of modern science was predicted by its major prophet, Sir Francis Bacon, in the seventeenth century. Besides being a major political figure and an essayist, he was also a major influence on the development of the new science, through *The Great Instauration*, a major discourse intended to announce a new beginning of human knowledge. Bacon never completed this work, but the basic fact is well-known that Bacon, along with his contemporary and correspondent Hobbes, attempted to counter the suppressive effects of Aristotelian philosophy in the universities of the time by constructing a new philosophy. Given the overall range of Aristotle's thought and writings, from logic to metaphysics to cosmology, Bacon and Hobbes attempted to replace the Aristotelian corpus point by point and field of knowledge by field of knowledge. The difficulty they faced, however, was that while it was barely possibly even for Aristotle in the fifth century B.C.E. to have broad knowledge of everything that was known in ancient days, enough new discoveries and methods had accumulated by the seventeenth century to make a like attempt simply futile. And rejecting Aristotle's complex and versatile metaphysics for a simplified philosophical materialism left the promoters of the new science without an intellectual means of developing a comprehensive vision.

Despite the metaphysical neglect, Bacon's writing revealed in a rough manner the prospect of what the major features of modern science would be. Modern science would be experimental; its general theories would be based on an accumulation of experimental and empirical data; the logical process by which theories were extracted from data would utilize the inductive method; and its explanations would reject teleology, divine intervention, and any other cause based on tradition and authority. These features are always credited to Bacon's vision, but another essential element of the new science, its social nature, is usually neglected. For Bacon, modern empirical science was primarily a social enterprise, necessitating at least a small population of experimenters, mathematicians, and theorists to collect and secure all the knowledge available by the scientific method in the various fields of mechanics, biology, astronomy, etc. Modern science for the first time was envisioned as the work of many lifetimes of many dedicated researchers. Bacon made this point not by means of a didactic essay, which was his usual method, but through a utopian fable, *The New Atlantis*. In the fable, a ship carrying European explorers lands on an unknown island inhabited by people of an advanced and irenic civilization who call their land *Bensalem* (presumably "good Salem," a reference to the Hebrew Bible.) Using such a utopian fantasy as a means of criticizing aspects of European society was not new, More had done it in his *Utopia*, which contained not only a utopian view of human nature and of society in general but also a sharp criticism of English policies such as the Enclosure Laws. However, Bacon's fable was not intended as social criticism but as prophecy.

The prophecy of Bacon was intended as a guide for the development of the new science that he and others were then espousing. In the fable, a spokesman explains to the Europeans, who had departed however from Peru, that there is in place on their island an institution or "foundation" called "Salomon's House." He defines its purpose thusly: "The end of our foundation is the knowledge of causes, and secret motions of things; and the enlarging of the bounds of human empire, to the effecting of all things possible."[11] Given its significance, this brief description needs to be unpacked. The reference to the *knowledge of causes* follows Aristotle's theory of knowledge pretty closely, while the reference to the *secret motions of things* is a general but accurate description of how scientific method approaches the task of the initial exploration of phenomena, that is, by looking to underlying entities and causes. The reference to *human empire* is significant because it refers not to a particular empire but to the whole effort of the human race to dominate nature (Bacon said that nature should be subject to torture to reveal its secrets). This is further indicated in the last phrase, which exposes in a naïve manner the whole aim of the new science: to attain a triumphant end of understanding and to control nature to the largest possible extent. Among Bacon's aphorisms is the phrase "Knowledge is Power." The new science will not lead to a fuller understanding of nature in a contemplative sense, that is, in terms of its final causes and relations to providence and to God as Creator, but rather to domination of the physical environment.

The description of Salomon's House that follows is detailed and takes up the last third of the *New Atlantis*; the first two-thirds comprises a narrative about how the discoverers came across the island, and their introduction to its people and culture. Salomon's House is divided up into several subordinate units, each of which conduct the separate types of research—including investigations of the Earth, the atmosphere, oceans, etc.—as Bacon then understood to be what scientists of his time were studying. At the end of this account, the spokesman describes the human resources that supply Salomon's House with its experts, theorists, technicians, and general manpower. There are nine groups of scientists consisting of three persons each, men and women, as Bacon puts it later on, the "employments and offices of our fellows." The functions of the nine groups reflect Bacon's idea of scientific method, a view that was neither premature nor contradicted by later developments. Thus, there is a unit that reads the results of prior experiments from books, as well as a unit that runs experiments itself. There is a unit of "Compilers," who divide the experimental results into "titles and tables" from which to draw "observations and axioms," which is a reflection of Bacon's reliance on induction as the means of obtaining general explanations from scientific experimentation. This would seem to be the basic feature, since the other units find and extend applications for the experimental results. The unit called "Interpreters of Nature" then carefully raises and systematically applies the experimental results, in multiple areas of investigation, into "greater observations, axioms and aphorisms." To ensure that the work of these nine master units will be sustained over generations, and that the knowledge so purposefully attained will not be lost, there is in place a system of training novices to succeed in these posts.

The entire enterprise depends on the work of a "great number of servants and attendants"—a small army, really. Bacon concludes his account with references to the means of publishing the results of the experimental research, and finally to a system of rituals and memorializations that reinforce among the citizens of Bensalem the great work of experimentation, theorization, and application that is the purpose of Salomon's House.[12] What is most important in Bacon's prophetic vision is that the progress of modern science requires a complex arrangement of the separate but interconnected parts of the large social organization that he predicts is necessary.

Not enough credit has been given to Bacon's perception of the social requirements and implications of modern science, which appears to have been entirely correct. Yet much criticism has been directed at his naïve, inductive philosophy of science. Bacon's vision has been derided by such major literary figures as Swift, who satirized the scientists of Salomon's House as nitwits who had to be reminded of practical reality by being hit in the head with a pig's bladder. Observing the impracticality of theoretical mankind is a criticism as old as the tale of Thales in the fifth century B.C.E. falling down a well while looking up into the heavens. The most savage characterization of Bacon's vision appears in C.S. Lewis's novel, *That Hideous Strength*, which was written during World War II when the consequences of organized scientific research as a means of social domination were devastatingly apparent. In the novel, Salomon's House becomes the National Institute for Experiments, or "NICE," and the collation of experimental results is accomplished by a Pragmatometer, a form of a calculating device. In Lewis's vision the aim of NICE is totalitarian.[13]

Science-Inspired Social Norms—Dewey's Progressivism

In modern political terms, Whig progress implies a liberal or progressive political ideology. Progress takes place in the scientific sense by means of new discoveries and advanced theories and such scientific progress is taken to imply progress for society at large, as scientific technologies relive the burdens of daily living. Most importantly, scientific ideals are employed as philosophical justification for social reform. The mechanical vision of the world picture was seen as a form of liberation from ancient forms of thought, including Aristotle's philosophy and the authority of the Christian Church in the seventeenth century; in effect, the progress of modern science meant the progress of human society. The new emphasis would not be conservative and attempt to preserve the most useful traditional forms and sensibilities, but instead seek to replace them with new, efficient forms and sensibilities bent toward utility. But elimination of traditionalist barriers to progressive reform meant also that, on an intellectual level, a new vision of philosophy would be required.

Dewey was a major figure in the reform movement in the United States that began as the Progressive movement in the early twentieth century. Thus, despite the passage of three centuries he is a prime representative of the connection between the Whig ideal of scientific progress and the political ideology of modern progressivism. Dewey's book *Reconstruction in Philosophy* was published in 1920, and was based on a series of lectures that he had been invited to give to an academic audience in Japan.[14] His reputation, which had reached international proportions by this time, was not, however, based on Dewey's reputation as a philosopher but as a social reformer. He had become known to an educated audience by starting an experimental school in Chicago, joining up with the early reform movement represented by Jane Addams.

Dewey's *Reconstruction* is a detailed exposition of his underlying philosophy of political and social reform. It is not enough to say that as a pragmatist philosopher, following in the footsteps of Peirce and James, that he wished to combine theory with practice, that is, the abstract and ideal concepts of philosophic discourse with practical applications that affected the lives of real people. Rather, as he explains in *Reconstruction*, Dewey wanted to reform philosophy in the manner of reducing its pretensions and making philosophy a motive for reform—and, it appears, little else. His view on the departure of modern science from the medieval world view is emphatic:

> The world in which philosophers once put their trust was a closed world, a world consisting internally of a limited number of fixed forms, and having definite boundaries externally. . . . a world of a limited number of classes, kinds, forms, distinct in quality . . . and arranged in a graded order of superiority and inferiority. [On the other hand] The world of modern science is an open world, a world varying indefinitely . . ."[15]

The idea that in the post-medieval world the options for mankind are literally infinite is characteristic of the Whig concept of progress, applying as well to its Enlightenment and Reductive variations, which will be explained in the following two chapters. It is a trope that is influential in the current age, as technological applications and the apparent spread of democracy happen worldwide and throughout every element of society. Dewey, however, is not replacing the medieval world order with an alternative description based on the discoveries of modern science, such as the Copernican theory; rather, for him—and reflective of the openness he perceives—what replaces medieval stasis is the *process of inquiry*. Dewey's concentration on process reflects the major theme of the philosophical movement of Pragmatism, as put forward by Peirce and James, namely, that of the necessary intertwining of abstract conceptual knowledge and practical application as they occur in human activity. Sometimes, however, Dewey's approach seems more reflective of Hegelian Idealism than of American Pragmatism.[16]

Dewey's reconstruction of philosophy is comprehensive; it is elaborated in separate chapters on: experience, logic, ethics, and social philosophy (chapters IV, VI, VII, and VIII, respectively). The verbosity and apparent evenhandedness of Dewey's writing, even as it reflects a good knowledge of historical circumstances, cannot disguise its radical intent. For Dewey, reconstruction in philosophy is a prelude to the reconstruction of society because it provides the intellectual and moral basis for progressive social reform. The culmination of Dewey's program appears in the final chapter, "Reconstruction as Affecting Social Philosophy." Dewey starts his analysis by setting the usual means of philosophical analysis of society in terms of the distinction between the individual and society, but then, in effect, dissolves the distinction to prepare for a more rational and useful discussion. Arguments and analyses based on this dichotomy are faulty because "[t]hey are general answers [that are] supposed to have a universal meaning that covers and dominates all particulars. Hence, they do not assist inquiry. They close it." Later on, Dewey states that the terms *individual* and *social* are "blanket terms for an immense variety of specific reactions, habits, dispositions and powers . . ." but that the "new method substitutes inquiry" into this multitude of facts.[17]

But where does this leave us, and Dewey, as we follow this presentation? Dewey's conclusion, unsurprisingly, is a new version of the sovereign state. But it is not the state as "a supreme end in itself" but rather as "an instrumentality for protecting and promoting other more voluntary forms of association."[18] The language has an indefiniteness about it as Dewey circles around subject matter that makes succinct summary virtually impossible. But, in the end, his desire is clear—not a state that asserts totalitarian supremacy and not individualism that leads to willful chaos, but something in between. Dewey relies on the process of inquiry to lead society into a more *integrated whole* (not a term that he uses, but accurate nonetheless). "Society is the process of associating in such ways that experiences, ideas, emotions, values are transmitted *and made common* (italics added).[19] Dewey's social philosophy is a kind of soft totalitarianism in which manipulation rather than outright coercion is the key to solving problems, conflicts, and contradictions as occur in society; here the basic pragmatic insight that intertwines ideas and actions is consolidated into the means of rearranging social patterns, processes, and attitudes.[20]

Dewey's social philosophy is an accurate representation of the aims of scientific progress as expressed in the fundamental Whig ideal of the scientific method of experiment, which leads to the discovery of general laws and then to the control of nature for the improvement of human society.

His social philosophy is also a good predictor of the kind of politics practiced currently by the democratic left in Europe and the United States, which favors "nudges" over ukases, regulations over laws, manipulation of public opinion over frank discussion of political issues. For Dewey, the reconstruction of philosophy means not only the reconstruction of society, but also the diminishment of philosophy as an independent study and as a source of high ideals, reduced to the handmaiden of progressive reform.

Notes

1. Herbert Butterfield, *The Whig Interpretation of History* (New York: Norton Library, 1965). Also, *The Origins of Modern Science* (New York: Free Press, 1957), Chapter 12, "Ideas of Progress and Ideas of Evolution, pp. 222–246.
2. Jean-Francois Lyotard, *The Postmodern Condition: A Report on Knowledge*; Minneapolis, University of Minnesota Press, 1984.
3. Gilbert Newton Lewis, *Valence* (New York, Dover, 1966), 35–54.
4. Loren Eiseley, *Darwin's Century and the Men Who Completed It* (Garden City, NY: Doubleday, 1958), 206.
5. Cf. S. J. Gould, *Rocks of Ages* (New York: Ballantine, 1999).
6. Galileo Galilei, *Dialogue Concerning the Two Chief World Systems*, trans. S. Drake (Berkeley, CA: University of California Press, 1967).
7. Stephen Toulmin, *The Return to Cosmology: Postmodern Science and the Theology of Nature* (Oakland: University of California Press, 1985).
8. Aristotle on causality, *Physics,* I-7, 8.
9. Lee Smolin, *The Trouble with Physics*; Boston, Houghton Miffline, 2006; x–xiii.
10. Amir D. Aczel, *The Jesuit and the Skull*; New York, Riverhead, 2007; 74–76.
11. Robert Maynard Hutchins, ed. *Great* Books of the Western *World*, Vol. 30, *Francis Bacon* (Chicago: University of Chicago Press, 1952), 210.
12. Ibid., 214.
13. C.S. Lewis, *That Hideous Strength*; New York, Macmillan, 1965.
14. John Dewey, *Reconstruction in Philosophy* (Boston: Beacon Press, 1957).
15. Ibid., 54–55.
16. Ibid., 134, 189–91.
17. Ibid., 199–200.
18. Ibid., 202ff.
19. Ibid., 207.
20. A.H. Somjee, *The Political Theory of John Dewey* (New York: Teachers College Press [Columbia University], 1968), "The Manipulative Aspect of Dewey's Political Theory" 160–178.

3

Enlightenment Progress and the Cosmopolitan Society

After a dissection of the Whig view of scientific progress and the allied concept of social progress, it might seem as if the general topic has been covered. However, there are discernible variations on the theme that are worth exploring. Among these are what is termed here the "enlightenment," view which places the subject in a more general pattern of civilizational development; the "reduction" view, which interprets progress in its materialistic form; and, finally a "historicist" variation, which tends to deny the reality of scientific progress and the concept of social progress. These are the subjects of the three subsequent chapters, which attempt a reinterpretation by introducing a more energetic philosophic interpretation of the whole topic of progress.

Enlightenment Variation of Scientific Progress

According to Immanuel Kant, its most profound representative, the Enlightenment was an assertion of mankind's independence from God in order to seek his own dignity and rejection of religion as a source of vital knowledge.[1] From this basic premise, certain consequences followed: first, that true enlightenment began in ancient Greece before the accession of the Christian faith to cultural predominance in Europe; second, that since true enlightenment had continued its existence independently of—albeit hidden by—intervening cultural and religious forces, it existed primarily as an *idea*; and, finally, that the idea of society that results from a suppression and recovery version of progress is one of cosmopolitanism, that is, of a universal city for all of mankind. The Enlightenment version of scientific progress closely follows these general themes.

The Enlightenment model of progress can also be termed the "suppression and revival" version, as an account of how Western civilization began with the ancient Greeks and progressed through Roman civilization, but then was suppressed the arrival of the Christian Church as the dominant cultural force in the West. This is the source of the term the "Dark Ages," made famous by Voltaire and other figures of the Enlightenment, which means that the culture of free thought, untrammeled speculation, and exploration originating in the high culture of ancient Greece was suppressed by the imposition of doctrinal orthodoxy, superstition, and bigotry. As an example, Cajori's *A History of Physics* (1929) states: "Obscurity and servility of thought, indistinctness of ideas, and mysticism characterize the Middle Ages. Writers on science were mainly commentators, and never thought of bringing the statements of ancient authors to the test of experiment."[2] With the arrival of the Enlightenment, Cajori implies, the tyranny of the Catholic Church was overthrown, which allowed the pure form of free thought, intellectual exploration, and, most relevantly here, the scientific spirit to arise again—or so the Enlightenment model asserts.

The Enlightenment was expressed in religious terms by the Protestant Reformation, in political terms by the rise of nationalism, and in intellectual terms by the rise of modern science. Modern science in this view is an extension of the thought of Greek thinkers such as Archimedes, Democritus, and Thales. This period was interrupted by the Dark Ages of church dominance in this view, a view of the history of science that implied that discoveries and theories such as Newton's would have been possible in the sixth century, but for the imposition of church heresy trials, as in the cases of Galileo, which ended in permanent house arrest, and of Giordano Bruno, which ended in death by fire.

As for the origins of science as a form of pure knowledge refined in the ancient fire of Greek culture, it is the opinion of a knowledgeable historian, Charles Gillespie, that ". . . science is a most arduous and unlikely undertaking. The answer lies in Greece. Ultimately science derives from the legacy of Greek philosophy." After discounting the practical techniques of the Egyptians and the Babylonians, Gillespie continues:

> Of all the triumphs of all the speculative triumphs of Greece, the most unexpected, the most truly novel, was precisely its rational conception of the cosmos as an orderly whole working by laws discoverable in thought. The Greek transition from myth to knowledge was the origin of science as of philosophy. Indeed, knowledge of nature formed part of philosophy until they parted company in the scientific revolution of the seventeenth century. . . . In our own world, science continues to be what it was in Greece, conceptual thought mediating between consciousness and nature."[3]

Many criticisms of the Enlightenment have appeared since the middle of the last century, when the above examples of science history were written, which amount to a general criticism that is reflected in contemporary versions of science history. Enlightenment thought was opposed in the eighteenth century by such figures as Blake and Burke, and by the authors and artists of the Romantic movement in the nineteenth century. But here we are concerned with contemporary critics of whom the most influential is probably Alasdair MacIntyre. MacIntyre is not a philosopher of science, but a well-known philosopher of ethics whose own reconstruction of the history of ethical thought led him to espouse a theory of *traditions* comparable to Kuhn's theory of paradigms (see chapter eight, where MacIntyre's social ethics are presented in more detail). In MacIntyre's terms, theories of ethics are not simply a concatenation of certain specified principles, but take form as lived entities that affect people's actual behavior and that has roots in the social context in which they live. In his best-known work, *After Virtue*, MacIntyre claims that after the collapse of medieval culture and the rise of Enlightenment thought, rational theories of social ethics such as utilitarianism arose as a matter of practical necessity. But in that transition, Western culture lost the ability to understand and state what virtue itself is.[4]

Enlightenment thought was based on the idea that there exists in nature one kind of reason that rejects bigotry and theology, and that arose independently of the religious constrictions of the Middle Ages; the rise of various philosophical ethical theories that competed with one another was the result. MacIntyre claims, however, that Enlightenment thought is not a universally applicable kind of reason—rational, all-pervading and self-secure—but in fact is a "tradition" no different in that sense from any other ethical tradition.[5] As a tradition and not merely an abstract philosophy, the Enlightenment idea was instantiated in the activities of its advocates during the French Revolution, as when a woman representing "Reason" was installed on the altar of the Notre Dame Cathedral in Paris after the religious images had been swept away. (The implication of the fact that the woman was a prostitute is not lost on observers, for Reason will also do what it is paid to do, as in the case of weapons scientists.)

In denigrating Enlightenment claims to universal reason, MacIntyre's thought resembles Kuhn's. Moreover, MacIntyre has acknowledged his debt to philosophers of science including Popper, Lakatos, and Polanyi, as well as Kuhn. MacIntyre and Kuhn effectively reset their respective fields from a positivist to a historicist form of philosophical understanding.[6] That the two most influential philosophers of science and ethics have opposed Enlightenment claims to universal rationality is a reflection of a general current in contemporary culture. Such criticism in general has influenced the critics of the Enlightenment version of scientific progress in two ways that will be explored here. First, how deep was the separation from medieval to modern scientific thought—was there truly a "scientific revolution"? And, since the Enlightenment's emphasis on rationality implies that the essence of science lies in the production of abstract theories, it can be asked, how correct a description of modern science is this?

Duhem's Continuity Challenge

The Enlightenment idea of a vast separation, a cultural transition of such depth that it is often called a "revolution," was first challenged in the late nineteenth century and not as a result of postmodernist ideas. The agent of challenge was Pierre Duhem, a late nineteenth- and early twentieth-century figure who was both a theoretical physicist and probably the most influential, and certainly the most interesting, of all the historians of science. The basis of Duhem's challenge to the idea of a revolution from medieval obscurantism to the era of modern science was based on his detailed search of documents from the thirteenth to the fifteenth centuries, which were available in the university libraries of France. Although the documents existed they had to be discovered in obscure places, and Duhem was often the first person to have read many of them for several centuries. Furthermore, the documents not only had to be unearthed, but had to be sympathetically and intelligently read so as to tease out their relevance to the advancement of *physical theory* (to use Duhem's phrase). It was to his advantage that Duhem could both read Latin, in which many of the documents were written, and was also a physicist fully conversant with all of that science's major theories and minor controversies.

When Duhem began his researches into the history of physics, he took for granted the Enlightenment view of the history of science. It was only by the force of the documents themselves that it may be said that he came to a view that directly challenged it. Of note here is that Duhem was a devout Roman Catholic at a time when anticlericalism was the policy of the French government, and it may be assumed that he was sympathetic to the idea that the Catholic Middle Ages had participated in the growth of modern science in a direct and positive manner. As Duhem dug into the documents, he reported each of his discoveries to well-known academic journals of the time; the general import of Duhem's researches was that the outstanding ideas and discoveries that characterize the scientific revolution had a steady stream of predecessors that led up to them. Duhem was able to trace the development of the idea of inertia in the work of Galileo from fourteenth- century figures such as Oresme and Buridan, who had developed the theory of *impetus*, which was a departure from Aristotle's qualitative theories of motion. Generalizing as both a historian and a theoretical physicist, Duhem constructed an overall theory about the nature of the history of physics in particular, and of science more generally—namely that of *continuity*.

> The historian, who is fond of simple and superficial views, celebrates the lightning discoveries which make the full daylight of truth to succeed the profound ignorance of night. But the one who subjects to a penetrating and detailed analysis the most novel and apparently most unexpected discovery, finds there almost invariably the result of a vast amount of imperceptible efforts and the confluence of an infinity of obscure trends. Each phase of the evolution which slowly leads science to its completion appears to that historian to be marked by two characteristics: continuity and complexity.[7]

Duhem's views on the development of early science contradicted the prevalent opinion of his time, a view that persists to a lesser degree in our own time. Duhem concentrated on the figure of Leonardo da Vinci as a major precursor to modern science, rather than as only an artist and sculptor of the high Italian Renaissance; instead Duhem took the observations and ideas that are contained in Leonardo's famous notebooks as serious contributions to scientific research. Despite the seeming oddness of this view, it allowed Duhem to give a clear portrayal of his major historical discovery that the great works of Galileo had predecessors among the despised medieval theologians who were investigating motion and statics in a quantitative manner. The treatment of Galileo was not meant by Duhem to diminish the importance of the chief figure in the rise of the modern scientific era; nonetheless, to show that modern science had not sprung fully developed from Galileo's mind forcefully contradicts a central belief of the Enlightenment view of scientific progress. The discovery of the continuity between medieval and modern scientific thought was, and—to some extent, still is—a major point of confrontation with the Enlightenment view, which endures and is often taken for granted.

Another contrarian view of Duhem's concerned the impact of the Condemnation of 1277, an episode in which the Bishop of Paris issued a condemnation of various heretical opinions fostered by Aristotle and his followers that was accompanied by a literal burning of heretical books. Among these were the works of Aquinas. Twentieth-century Catholic philosophers and historians of philosophy, chiefly neo-Thomists have often denounced the Condemnation because its effect was to constrict the intellectual field in which medieval scholars had to work from that point on, given the exclusion of Aristotelian doctrines. For neo-Thomists such as Gilson, the Condemnation of 1277 meant that Aquinas's fecund combination of Aristotle's philosophy with the doctrines of the Christian Church could no longer be developed.[8] In the view of Duhem, however—a historian of physics rather than of philosophy—the Condemnation had the positive effect of releasing scholars whose interest was in natural phenomena rather than theology from Aristotelian theories. These included the qualitative theory of motion, the sharp differentiation between earthly and heavenly phenomena, and the cosmology of the heavenly spheres. Relying on the Christian teaching of God's omnipotence, scholars of late medieval times could speculate that the divine Creator could build the universe in an instant, giving motion to it and to all its parts that would last to eternity; this teaching was eventually connected to the principle of inertia.

A problem for the interested observer in seeking out Duhem's historical works is that they are not readily available in English, having largely been left untranslated from the original French. Even in French, much of his historical work has not been republished or edited; indeed, the later volumes of his massive history of cosmological theories have not even been published. This is a loss as students are not able to see for themselves the quality and detail as well as the means of interpretation of Duhem's historical efforts. There is in print a selection of Duhem's historical writings, but it consists only of brief texts. However, a detailed summary of his historical effort in physics is readily available online. Duhem wrote a sixty-seven-page article bearing the title "History of Physics" that appeared in the *Catholic Encyclopedia*. Even though Duhem's history does not include twentieth-century developments, its usefulness is indicated by its placement on a well-known search engine, in which the search for "history of physics" shows that Duhem's account is the second- to-fourth most-read article, right after the Wikipedia treatment or Asimov's history.[9]

Duhem's Philosophy of Science and the Continuity Thesis

Duhem was as much a philosopher as a historian of science, and his views on both fields of study influenced the other. He developed a fairly coherent philosophy of science, which, while written in the early part of the twentieth century, anticipated more recent developments, mainly those that have become prominent with the decline of positivism as the chief philosophical underpinning of the philosophy of science. There are elements that are similar to the points that Kuhn, Popper, and others subsequently made about the impermanence of theories and their rather more fragile connection to a presumed underlying physical reality than is assumed by most philosophers and practitioners of modern science. Duhem's historical researches had shown him that physical theory advances in small steps as its elements are accepted, corrected, or rejected over time. Theory is affected by the consideration of experimental evidence, and is altered by its relationship to philosophical strands of thought as well. The idea that a given physical theory attains a final form is a mere expectation in that theories are not the final conclusions of inductive reasoning; the final form of a theory does not remain unaltered, and there are no "final" or perfected theories in science—a lesson that Duhem learned from studying its history. He may have learned this indirectness of scientific theory not only from a study of its history, however, but also by his own acquaintance with the state of physics in his time. By the late nineteenth century, while relativity and quantum theory had not yet revolutionized physics, it was obvious to theoretical physicists that the Newtonian mechanical synthesis was no longer adequate to explain new discoveries in radioactivity, fluid mechanics, electromagnetism, and atomic physics.

The very granularity of the advance of physics, as Duhem understood it, presumed against a purely theoretical interpretation of the history of physics and of science generally, "theoretical" in the sense that theory was superior to empirical discovery, so that the formation of general theories is assumed to be the true end of science, as exemplified by Newton's *Principia*. The Enlightenment view of scientific progress assumes usually that theory, not practice or empirical observation, is the true essence of science. But what then is the status of physical theories? In particular, what is the relationship between theories and the physical reality that they are supposed to explain? To ask those questions, however, is to leave the realm of science as a research activity and to enter into the realm of metaphysics; indeed, Duhem begins his book on the philosophy of science by asserting that ". . . the value of a physical theory depends on the metaphysical system one adopts."[10] He adduces examples from Aristotle, Newton, and Descartes. Such a view is positivistic in the sense that it reduces theoretical explanations to formal constructs that cover the experimental evidence. This implies that theories will be successively altered as time goes by as additional empirical discoveries are made in that particular field of scientific research; again, no "final" theory is presumed to lie at the end of the process of discovery. According to this view, theories have a largely functional existence, for their abstract formalisms do not have direct references to empirical evidence but exist rather as symbolic relations.[11] Here Duhem revives a medieval term, to "save the appearances," with the understanding being that formal theoretical constructs have no actual ability to describe reality in itself. Duhem's final philosophy of science opposes inductivism as exemplified by Newton, and mechanism as the general method and the underlying metaphysics of physics.

Duhem's philosophy of physics is connected to his historical perspective in two ways. First, that theories have only an indirect relation to the physical reality they are designed to describe comports well with his view that progress in science happens in small, granular episodes, the result of which is to flatten the assumption that science proceeds in grand, dramatic stages. Such stages are typically seen as the announcement of a singular, new, and comprehensive theory; Copernicus's *De Revolutionibus*, Galileo's Law of Falling Bodies, Newton's *Principia*, Darwin's *Origin of Species*, and Einstein's Theory of Relativity provide the best-known examples. In such a presentation, however, the work of theoretical predecessors, and of technicians and experimentalists, is often ignored. The second connection of continuity theory to Duhem's view of science history is that it deflates to a large degree the perception that there is a great gulf, intellectually and socially, between the medieval worldview and the modern. There is, Duhem asserts, a continuous path from the researches of the late medieval scholars who were examining Aristotle's theory of motion and developing the mathematics that distinguished between steady motion, accelerated motion, and motion that accelerated at an increasing rate. These distinctions were predecessor developments that led up to the discoveries of Galileo and of Descartes.

The granularity of scientific progress also implied for Duhem that there are no such things as "crucial experiments," that is, those that can eliminate or make certain a particular hypothesis.[12] This idea was taken up by the American logician Quine and is now known as the "Quine-Duhem" thesis. In Quine's updating of it, however, the thesis is really an epistemological rejoinder to positivistic assertions that the meaning of a term can be understood with definite clarity apart from its context within a text. Quine's point was that the meanings of terms depend more often on their relation to other, related terms.[13] In either version, however, the rejection of crucial experiments contradicts the Enlightenment model of science history as the process of inevitable progress leading to correct and culminating theories.

Cultural and Social Criticism

The major opposition to the Enlightenment version of progress arguably comes from historical researches done in the last twenty years or so, historical works that often reveal an ideological edge. The contrast exists between the Enlightenment's high regard for ideas in general and for scientific theories in particular on one hand, and in the current concentration on the granular, empirical side of science that characterizes science history as an accretion of empirical discoveries over time. This kind of history is not limited to details of research, but makes connections with broader cultural developments as well. The understanding is that science is a human activity emplaced inevitably in a web of interests and influences that are cultural, political, economic, and religious. This oppositional view has philosophic implications for the relative importance of empirical research and for experimentation as the primary essential aspect of science as opposed to the development of abstract theories, no matter how well-supported by experimental implications. This antithetical perspective often has a left-wing ideological edge to it, since it does not concentrate on the large theories and great scientists usually presumed to constitute the top of the pyramid of scientific history. Instead, this view focuses on the base of the pyramid, amid the struggling workers who leave no record for posterity, those "miners, midwives, and low mechanics" who are neglected and who remain anonymous in the usual scientific histories. Such accounts give credit as well to the often-unrecognized research done by women scientists.

An interesting and provocative example of left-wing science history is Clifford D. Connor's *A People's History of Science* (2005). "People's histories" have become somewhat common, including, for example, a "people's history of art," all of them portraying their areas in terms of the many workers whose supportive labor has gone unrecognized; the paradigmatic example is Howard Zinn's *A People's History of the United States* (1980). A people's history of science may seem not to easily fit this kind of history at first glance, but since empirical science is based upon observation and research, and since such research must be done by a myriad of people who are knowledgeable in various crafts such as glass blowing, electrical wiring, and metallurgy (and, these days, accountants, office assistants, and grant-applicant specialists), Connor easily makes his case that modern science is a social enterprise.

Among assertions that contradict the Enlightenment view is Connor's explanation of the death of scientific progress in ancient Greece, which he attributes to Plato's philosophy. The high abstractions of this philosophy are the result of the leisure available to Athenian upper classes, because the unremitting labor required for such a polite existence was provided by slaves. There was no impetus for gradual progress in agriculture, for example, since greater efficiency was not required for greater productivity given the ready availability of slave labor.[14] Connor also cites the example of Boyle, who is credited with the discovery of laws that explain the behavior of gasses. He contravenes the statement of historian Marie Boas Hall that "the names of Boyle's assistants have been lost, because they were not men of independent scientific merit," citing the research of Shapin, which points out that Boyle was a wealthy man who could afford to employ several full-time laboratory assistants. Connor notes other such examples, including Lord Rutherford and J.J. Thompson.[15]

While Connor's point of view is social and ideological, an emphasis on the cultural as opposed to the theoretical side of modern science also has implications for the philosophy of science. That is, the relationship between a scientific theory and the empirical evidence for it, as well as the cultural and social contexts in which it exists, must be seen as more intimate. A theory is not to be seen in Platonic fashion as an abstract entity, the reality of which is superior metaphysically to its material exemplifications, or the expression of which by intellectuals must be socially superior to those of the *hoi polloi*. The empirical research, including the technical practice and material implements used by laboratory technicians and experimentalists, gives a flavor to the theory and defines it pragmatically rather than abstractly, it may be said. There is no gulf separating the empirical side of science from the theoretical, a more common realization these days than formerly. This results in the identification of science with the activity of researchers as much as of theorists, a point that has egalitarian implications.

Among current historians of science with a philosophic bent, Peter Galison has expressed this anti-Enlightenment philosophy energetically, clearly, and somewhat elegantly. He has introduced the term "trading zones" to indicate a space wherein the three important aspects of scientific explanations, philosophical speculation, and cultural memes can interact and mutually impact each other. Galison wrote a major work that expresses this philosophy by means of an analysis of what happens "on the shop floor" in regard to the discoveries that led to the theories that violated the canons of classical scientific explanation, but that nonetheless are the basis of the extraordinary development of quantum physics.[16] Among Galison's many efforts was editing a collection of essays written by several recognized authors with the provocative title, *The Disunity of Science*.[17] Galison's view powerfully resonates with the contemporary egalitarian *zeitgeist* that refuses to acknowledge any difference between high and low culture. Galison's energetic endeavors in this line have attracted much attention, earning him a MacArthur Grant and a named professorship at Harvard.

The major work that emerged since Galison has been at Harvard is his volume about the origins of the theory of relativity, entitled *Einstein's Clocks, Poincare's Maps*.[18] It is a compact book of over 300 pages, intensely researched and neatly packaged; he notes the several foreign libraries where he has looked up original documents and names the many graduate assistants, editors, and colleagues who helped him write and produce the volume. Apparently, the work is intended as, if not a culminating, a clearly representative expression of Galison's intellectual approach. The book's subject is how relativity theory came about, with particular reference to a technological application that was of intense interest at the end of the nineteenth century, when railroad travel was extending all over the European continent and when timing of the departure and the arrival of fast-moving trains was essential to organizing this vast, new system of transportation.

The key was the use of electrically powered clocks that would keep the most accurate time that was then technologically possible. As important and of the most significance, however, is that the electric clocks had to be coordinated from a central location to other locations, so that what Galison refers to as "empires of time" was a universal phenomenon that depended on the fact that time itself could be measured within tenths of seconds. But in attempting to coordinate times at large distances, the fact that electricity traveled along wires at the speed of light was a complicating factor that had to be taken into account. As a practical and theoretical matter, it seemed as if absolute simultaneity was impossible. Galison recounts in detail Einstein's work at the Bern patent office, which concerned many types of patents for electrical transmission and clocks. In addition, Einstein's famous thought experiments used examples from railroad travel with reference to the speed of light: "Patent technology and theoretical understanding were closer than they might have appeared."[19]

During the same period, another theoretical physicist and mathematician, Henri Poincare, was investigating the same subject from a practical point of view. He was engaged by the French government to help resolve issues of unification and standardization of measurements of such things as longitude, distance, and, relevantly, time. Poincare extended his detailed knowledge of the problems involving electrical transmission of time measurement and the impossibility of achieving simultaneous time recognition to the theoretical level, which led him to dismiss the ideal notion of a universal time such as Newton's absolute time. The burden of Galison's book is that, in effect, special relativity was a case of simultaneous discovery, proving that the theory arose from the practical necessity and the technology of the coordination of electrical clocks in different locations by means of electrical impulses traveling along copper wires at the speed of light.

Galison's case is flawed since Poincare's theoretical ideas, while paralleling Einstein's, do so only up to a point. It was Einstein and not Poincare who made the speed of light a constant and who worked it through his equation. In addition, it was Einstein's theory that made it possible to explain electromagnetism and classical mechanics by a single set of principles, and that provided the basis for a mathematical description of space-time. Actually, it is the discovery of general, not special, relativity that provides a clear case of simultaneous discovery, when the mathematician Hilbert and Einstein were vying against each to be the first to invent the complex set of equations that permanently defined relativity by including gravity. But this was a race that involved developing abstract mathematical equations, an almost purely theoretical activity with only an indirect relationship to the work on the "shop floor." Developing general relativity was, at that point in time, a matter of head work not hand work, a fact that does not comport with Galison's ideas about the technological basis of the scientific enterprise.

Koyré's Infinity Defense of Enlighted Science

Given the historical examples of medieval science researched by Duhem, as well as the newly developed cultural and social descriptions of science history that denies that the invention of theories is the pinnacle of scientific production, what remains of the Enlightenment view of science? Quite a lot, in fact, as the unity of science throughout its history remains one of the prominent aims of theorists, for instance, Hawking and Weinberg, and as long as it does, the Enlightenment model will retain most of its plausibility. A defense has been made, albeit one that is out of historical sequence for it was written not in response to Galison and the recent school of the social history of science, but apparently in direct response to Duhem's continuity thesis. The work is Alexandre Koyré's *From the Closed World to the Infinite* Universe—published in 1957 and still in print—a summary of two of his lectures and previous scholarship.[20] It was his aim to reestablish the idea of a severe rupture in European thought, a "deep revolution" that involved not only scientific knowledge, but also philosophy and even theology. This was in contrast to Duhem's continuity view that denies that the rupture took place, or at least did not do so anywhere near the degree as is presented in the Whig model.

Koyré's history of science is of a distinct kind for it is based on a history of ideas, and does not often refer to empirical evidence or to experimental results as having an influence on the outcomes of large physical theories. His presentation of Newton's theories, which extends over three chapters, is not based on an analysis of the *Principia* regarding, for example, the three laws of motion or the tides, etc. Rather, the expositions concentrate solely on the philosophical issues from the *Scholia*, appended by Newton to his *Opticks*, on the nature of space and time, the nature of divine action implicit in Newton's account, as well as the controversy with the scholars of the Cartesian school over the definition of mechanical philosophy.[21] Koyré's primary concentration on the ideational rather than the empirical suffuses his historical approach in order to demonstrate, in effect, the unity of science—from its origins in ancient Greece to its reappearance in fuller bloom at the end of the medieval period. The basis of that unity is the idea that the physical universe in infinite.

Koyré traces that idea from its ancient roots in the cosmological speculations of Greek and Roman authors to the late medieval period in the fourteenth century, when a nominalist philosophy had replaced the theological metaphysics of the high medieval period of the thirteenth century. In the middle of the twentieth century, Neo-Thomists tended to see Nominalism as a result of the decay of the Christian philosophy of the middle ages, which reflects a split between those who study the progress of scientific thought and those who study the history of Christian philosophy. The theological school of Nominalism led to emphasis on the will of God rather than, as Aquinas had largely tended to think, on the intellect of God. For later philosophers and theologians, this meant that it was possible to speculate that God in his infinite ability to create had created an infinite universe.[22] This universe was infinite not only in its extent, but also might contain from the moment of creation all the initial forces that would continue to operate throughout its history. This idea of the *vis inertiae* culminated in the development of the modern theory of inertia that is essential to Newtonian mechanics.[23] In this manner, the mechanical theory that underlies classical mechanics has in part, historically speaking, a theological origin.

In a 2014 television program, a "reboot" of Carl Sagan's "Cosmos," the tragically persistent Giordano Bruno was featured as the main figure who, in the seventeenth century, introduced into science the idea of the infinite universe. Bruno, now an iconic martyr of the freedom of thought, conflated the infinity of God with the infinity of the universe in a manner that led him to explicitly deny the Trinity and the divinity of Christ, the two most essential doctrines of Catholic faith. This act culminated in Bruno's trial and gruesome punishment. In Koyré's account, which is much more thoroughly researched than the tale presented in the television program, the main character is Nicholas of Cusa, a fourteenth-century figure presented by the historian Gilson, as the last important person in the history of medieval philosophy, whose work portended its "breakdown."[24] Nicholas, not Bruno, is portrayed by Koyré, as the main carrier of the idea of the infinity of the universe to subsequent theorists.[25] Another prominent figure in Koyré's estimation is the obscure figure of Thomas Digges, an English astronomer whose diagram of the solar system is reproduced in *From the Closed World*.[26] Digges's diagram is almost indistinguishable from any other such picture of the Copernican system, for while the sun is at the center and the planets circle around it, what is of importance for Koyré is what is portrayed at the outer limit of the solar system. There is no circle or sphere of fixed stars giving a boundary to the solar system, but instead stars are randomly placed so as to represent that the universe in which the solar system exists is infinite. From this point on, the concept of the universe has radically changed, for the relative positions of the Earth and the sun are of less importance in a cultural sense than is the fact that the physical universe has no bounds.

It is too much to ask of any historian that he or she extend such an account as Koyré's, which covers a vast extent of time and still remains of interest fifty years after its publication, for an additional three centuries. Nonetheless, it is unfortunate that Koyré did not consider the concept of physical infinity implied by the general Theory of Relativity. The relativistic universe is infinite only in that it is unbounded (existing in four dimensions). But its mass, however vast, is nonetheless finite such that its value has been estimated, as has even the entire number of the universe's protons.[27] Further, the space that fills the relativistic universe is not the empty and homogenous space of Newton's that terrified Pascal. Rather, it has been transformed—filled with gravity waves, black holes, dark energy, and clouds of radioactive gas—and is "bent" by massy objects such as stars and large planets.

In pursuing such a vast stretch of history, one that extends from ancient times up until the seventeenth century (a span of 2,200 years), and given also that Koyré is writing a history of ideas and not of technological developments, the question of how to trace influence from one thinker to another becomes an issue (as it was for Duhem.)[28] Koyré is not writing the history of science in modern times, when influence can easily be traced in references to prior experimentation in a range of scientific journals. And Nicholas of Cusa, for example, did not specifically mention Lucretius or other ancient thinkers whose ideas he may have taken up and developed.

Rather, the question of influence is measured by the resemblance of separate ideas to one another, an evaluation that can become somewhat complicated. Koyré, however, relies on one fundamental idea—the infinity of the physical universe—that is so clear and overpoweringly significant that it would seem relatively easy to trace and to identify from thinker to thinker. As Koyré presents his case, the idea of physical infinity has rich resonances, linking as it does science, philosophy, and "even theology," effectively transforming the European mindset and resulting in the "utter devalorization of being, the divorce of the world of value and the world of fact."[29] Physical infinity is the thing that unifies ancient and modern thought, and its suppression, it is implied is responsible for the crabbed theologies and limited mindset of medieval culture. It was by this manner of explanation that Koyré defended the Enlightenment concept of suppression and revival against the historical attacks of Duhem and, implicitly, of Galison and of Connor.

Causality and Mechanical Reason

The basic issue of Enlightenment progress is the question of what was left behind from ancient Greek thought that now needed to be revived in order for European society to advance. According to Koyré, it was the concept of physical infinity, but surely there was more to it than that one idea. According to the historian Gillespie, quoted at the beginning of this chapter, what was suppressed was the concept of reason itself as a mediator between the abstract and practical everyday life. Indeed, in the broadest terms of popular presentations and rhetoric, the contrast between the ancient Greek and medieval cultures is that between unfettered reason and religious bigotry. What the Enlightenment revived was the idea of unbiased but effective reason itself, which could now be taken down from high levels of philosophic abstraction and applied to discover the laws by which the physical universe runs, and subsequently to assist mankind by means of scientific technology. But *this was not so much a revival of the ancient ideal as a reimagining* of it, for in hindsight what the Enlightenment thinkers thought they perceived in Greek culture was in fact a reflection of their own assumptions about the constitution of reason regarding the relation of human thought to the universe.

Bacon's philosophy of science was a kind of anti-philosophy, for Bacon specifically repudiated the philosophy not only of Aristotle, but philosophy (and theology) in general. The usual example referred to is the emphasis by Bacon on inductive as opposed to deductive reasoning.

(Aristotle however appreciated both forms of reasoning and described their interaction in his *Prior Analytics*). More important in the transition from the medieval to the modern scientific viewpoint was the rejection of Aristotle's teaching on causality. Causality was the first casualty of Bacon's rejection of philosophy, and the most obvious factor in that rejection is that while Aristotle's doctrine of causality is of multiple types, that of modern science is of one type only. That is, Aristotle described four kinds or types of causes: formal, final, efficient, and material. In the making of a house or a bed (Aristotle's examples were always homey), the formal cause is the design or the blueprint of the thing being made; the final cause is the purpose of the artifact, for instance, a place to sleep or live in; the efficient cause is the actual maker, the carpenter or the construction workers; and the material cause is the "stuff" from which the artifact is made such as lumber, stone, cement, and iron nails. In the rejection of formal and of final causes especially, the Baconian viewpoint virtually assured that a materialistic description of reality would prevail, for the elements of mind and purpose were also casualties of the rejection of Aristotelian philosophy. Thus, it is notable that scientific opponents of biological determinism in our own time, such as Lewontin, have found it useful to reintroduce the idea of multiple modes of causality.[30]

Bacon's rejection of final and formal causes resulted in a mechanical description of causality and, ultimately, of reason—a rejection of the innate idealism and comprehensive rationalism that was thought to involve not only men's minds but also the constitution of the universe itself, a metaphysical insight that originates in Plato, Aristotle, and the Stoics. It is just this concept of reason as a nonmechanical and penetrating aspect of all reality, subjective as well as objective, as constitutive of universal reality that has been excluded by the Enlightenment view. What we may call this "comprehensive" view of reason was finally eliminated from much Western science by the acceptance of the mechanical view of the universe. Ironically, the attenuated view of scientific reason is reflected in the fact that modern science has never really been able to define a concept of causality that is stable, or permanent. Hume, in the eighteenth century, described scientific causality in terms of the interactive clicking of billiard balls. But Newton was able to apply the calculus to the actions of billiard balls, as well as planets and all other examples of material bodies, giving his scientific approach great scope and accuracy. As a consequence, we may conclude that the effective model of scientific mono-causality has two essential characteristics: material and mathematical. In Aristotelian terminology, it combines material and formal types of cause, but given its mathematical aspect it is really a wholly different concept of causality altogether.

By the late nineteenth century, the deterministic calculus would be of much less use in describing physical phenomena that are more complex than planets or falling bodies, for the behavior of gases and liquids would require the use of probability mathematics. And here the notion of scientific causality would have to change from a deterministic to a probabilistic mode, which raised a serious problem. That is, in cases of molecular interaction, causality was now based on the correlation of two separate phenomena, for example, an increase in heat applied and an increase in temperature, thus implying cause and effect. But was such a linkage mere circumstance, the result of chance, or did it indicate a real connection between events? The atomic theory of heat would provide the link. However, scientists who examine the link between cigarette smoking and cancer, between wine-drinking and long-term health, or between single parenting and the patterns of development of offspring would be challenged to examine the correlations to see if they were real or not. Nor is the problem necessarily easier for the hard sciences, for atomic physics does not escape the necessity of using probability, as is well-known in quantum phenomena. But at this historical point, the concept of scientific causality that is based on both materialism and mathematical prediction has been significantly weakened as a useful concept. Atomic materialism has dissolved from hard, impenetrable atoms to a whirl of subatomic forces and evanescent particles, while the mathematics have devolved from differential calculus (deterministic) to regression analysis (probabilistic).

Kant on Diminished Reason and the Cosmopolitan Social Ideal

What then is the implication for an ideal society of scientific causality and reason? The great expositor of this question was, of course, Kant, whose most famous works were on reason, both pure and practical; in both cases, he wrote *critiques*, that is, critical appraisals. The reduction of reason from a comprehensive form, as in ancient and medieval philosophy, to a mostly utilitarian functionality as the result of the rise of modern science is the underlying premise of Kant's thought and, consequently, of his social philosophy; Kant discerns quite clearly the difficulties inherent in the concept of reason he has inherited. Kant wrote about ethics as well as scientific reason, and developed a complicated but compelling theory by which he claimed the morality of all deliberate human actions could be determined *a priori*, that is, not based on experience, religion, or philosophy, and, further, without consideration of its practical effects (by means of interior ratiocination [reason] alone). In this way, argued Kant, we attain the purest region of unselfish and unbiased decisions. All morality could be determined by one categorical imperative: "Therefore there is only one categorical imperative, namely this: *Act only on a maxim by which you can will that it, at the same time, should become a general law.*"[31] For Kant, there is nothing moral in human actions unless it is directed by a person's good will. This leaves a central objection to Kantian ethics: if a person is "sincere," then no matter how objectionable or immoral his action may be, the action is a good one.

The usual assumption of undergraduate students, when first encountering ethical theory, is that if Hitler was sincere and really thought he was doing what was best, that is, on the maxim that he would will that his beliefs become a general law, then his actions were moral! This seemingly jejune response is seemingly unanswerable on Kant's own terms, which explicitly deny that the consequences of an action are not to be considered in evaluating its morality. Kant, however, is among our greatest philosophers, which is seen in his further doctrine that human individuals are to be treated as *ends* not *means.* Indeed, it is this doctrine that renders Kant's deontological ethics an essential part of theoretical discourse these days; however, it is difficult to see how his doctrine of evaluating the morality of actions in terms of a good will alone is connected in an inherent manner to his pronouncement of the sacred status (to use alternate terminology) of the individual human person.

But, again, what is the consequence of the Enlightenment model of science history for an ideal of society? According to Kant, the ancient ideal now revived was that of *cosmopolitanism*, the ideal that by means of human reason alone, mankind could aspire to a universal city of peace, cooperation, and comity in which every person was a citizen, with no distinction based on class, gender, race, or religion. Kant was not a historically minded philosopher; all of his philosophizing seems to take place on a plane of abstraction that depends, in his expression, "on reason alone." This means that for Kant, references to the political upheavals, wars of religion, and intellectual ferment taking place in his time were not only not mentioned, but not even referenced indirectly. Thus, understanding Kant depends on taking the method he developed at his word, that it depends on the use of abstract reason; that methodology, however, leads to an indeterminism and an obscurity in the ethical and social philosophy of Kant. Nevertheless, he wrote several essays on ethics and on social philosophy, including "What is Enlightenment," "Eternal Peace," and "Religion within the Limits of Reason Alone." In what follows, his most relevant essay, "Idea for a Universal History with Cosmopolitan Intent," will be briefly analyzed.

That Kant would write any kind of a history in the manner of a linear description of actual events seems unlikely given his penchant for abstract reasoning, independent of individual circumstances. Typically, then, Kant's essay is not a universal history but an idea for such a thing, "Thus (it is to be hoped) that what appears to be complicated and accidental in individuals, may yet be understood as a steady, progressive though slow, evolution of the original endowment of the entire species." As for the actual substance of what constitutes the history of humanity, Kant states frankly, "That it is hard to suppress a certain disgust when contemplating men's actions upon the stage." Nonetheless, Kant states that he will soldier on to "see whether we can succeed in discovering a guide to such a history."[32] In the manner of a conscientious college teacher, he lists nine principles by which such a history should be written. Given its idealized nature, the eventual link of Kant's treatment to Hegel's historicist philosophy is evident, for Hegel did attempt to write "universal" histories.

The principles, which will not be listed here but only summarized, start by stating in the most general way that "All natural faculties of a creature are destined to unfold . . . according to their end," and that in man, since he is the only rational creature, it is the faculty of reason that will be fulfilled, *not in the individual but in the species.*[33] Kant's conflicted attitude toward reason is reflected here. It seems to combine a nearly Platonic assumption of reason's power, along with contempt for the human beings who are the bearers of this exalted power. Nevertheless, Kant states that it is up to man himself to transcend the "mechanical ordering of his animal existence," in order to self-create his own nature independent of his animal nature, as it were; this is intended by nature itself. This is stated in Kant's Third Principle, which seems in a way to be self-contradictory, for man is supposed to somehow transcend nature (mechanical ordering?) by an act of collective self-creation (as per the Second Principle), a process that is intended by nature itself. All this is not some alien speculation but the current basis of Progressive thinking, that is, the idea that there is a vast plan of historical progress intended by nature, which applies specifically to the human race which must transcend nature by its own powers. This phenomenon, however, is seen in its generality rather than in human particularity.

In principles four, five, and six of cosmopolitan history, we discover that mankind's anfractuosity needs to be overcome by the establishment of civil society. This would be accomplished by means of a general civil law (hence, the basis for a Kantian version of the contract theory of society). But further, given the mutual antagonism of individuals against each other, which is always present and is the overriding feature of human relations, according to Kant, we face "the most difficult" problem, as addressed in the Sixth Principle. "Man is an animal who, if he lives among others of his kind, needs a master."[34] For even though man is a rational being who acknowledges the need for civil law, the egoistic inclination is to exclude himself from that law whenever he can. But now man faces a dilemma that Kant acknowledges can never be fully overcome: if man needs a master, who else can that be but another man who contains within him or herself the antagonism against others of his or her kind, and the innate inclination to exclude him or herself from the force of law even as it is supposed to be universally applied. Thus, although up to this point in his analysis Kant seems to be directly reflecting the thought of Hobbes with his version of the war of all against all, Kant is aware of the faulty nature of the tyrant who Hobbes claims is the means by which civil society can be successfully established. Kant is aware that the King or Protector must be a man himself, and the acquisition of such ultimate power in one man is obviously dangerous. Kant faces the dilemma that Hobbes does not—anarchy or tyranny, King Log or King Stork. What then is Kant's solution?

In the final three principles, Kant elaborates his vision of a natural progression of human history to a cosmopolitan society. The basic means for this intended end of progress is the state (Seventh Principle) under the realm of a constitution, indeed a "perfect constitution" (Eighth Principle).[35] However, the state does not exist independently but in a coalition of like-minded states. In this manner Kant's cosmopolitan vision is an international one, since a combination or alliance of all the best developed states, that is, those that have traveled the furthest along nature's designated path to progress, will result in a world of freedom under law. Interestingly, this is not his last word on the topic. Kant reflects back on his vision to promote the initial idea of his essay, that "a philosophical attempt to write a general world history according to a plan of nature which aims at a perfect civic association" must be considered possible and "even helpful" to attaining that very end. That is, writing a history that recognizes and describes the progressive historical process now under way itself contributes to the progressive intention of nature

(Ninth Principle).[36] It is this idea of history wrapping around itself so to speak, that will come to fruition in Hegel, for whom in a rough sense the *idea* of history is "real" history.

What Kant has accomplished in his essay is the first complete description of the progressive, cosmopolitan vision that is the social aspect of the Enlightenment idea of progress. In it, we see what contemporary progressivism views as the destined history of mankind: reaching toward a society in which all men are brothers/sisters, not by dint of a common human nature based on biology or psychology, but on reason alone. So ultimately abstract is this vision, both in Kant's presentation of it and in current manifestations of progressive politics, that individual human beings possess no personal, national, ethnic, or religious histories, but are seen as ciphers characterized only by their possession of (imperfect) reason. They are expected to be able to ignore their own personal circumstances and even supposed to be able ignore the most pressing self-interest, in order to gain an "objective" view that is based on unbiased reason (a remote objectivity guaranteed by Rawl's famous "veil of ignorance"). Yet, in Kant's view, reason itself is what is left of the ancient ideal of human reason after modern scientific philosophy has denuded it of its philosophic reach. Therefore, aside from this general idea of progress, according to Kant reason cannot positively comprehend ethical values, the design of the cosmos, or the existence of God. And this must be understood as a great loss.

Notes

1. Immanuel Kant, "What is Enlightenment," *The Philosophy of Kant*, ed. C.J. Friedrich (New York: Modern Library, 1949); 132.
2. F. Cajori, *A History of Physics* (New York, Dover, 1929); 24.
3. Charles C. Gillespie, *The Edge of Objectivity* (Princeton: Princeton University Press, 1960) 9.
4. Alasdair MacIntyre, *After Virtue* (Notre Dame: University of Notre Dame Press, 1984).
5. MacIntyre, *Whose Reason, What Rationality* (Duckworth, 1988).
6. See John Caiazza, "The Influence of Philosophy of Science on MacIntyre's Ethical Thought"; *American Catholic Philosophical Quarterly*; Fall, 2014; Vol. 88, No.4.
7. Cited in Stanley L. Jaki, *Uneasy Genius: The Life and Work of Pierre Duhem* (the Hague, Netherlands: Nijhoff, 1984) 388–89.
8. Etienne Gilson, *History of Philosophy in the Middle Ages* (New York, Random House, 1955), 408.
9. http:// newadvenst.org / cathen/12047a.htm.
10. Pierre Duhem, *The Aim and Structure of Physical Theory,* trans P. Weiner (Princeton: Princeton University Press, 1982), 10–14.
11. Ibid., 27, 79.
12. Ibid., 188–90.

13. Willard Van Orman Quine, *From A Logical Point of View* (Cambridge, Harvard University Press, 1961), in "Two Dogma's of Empiricism", 41.
14. Clifford D. Connor, *A People's History of Science: Miners, Midwives, and "Low Mechanics,"* (New York: Nation Books, 2005); "Plato's Elitism," 142–44.
15. Ibid., 335–36.
16. Peter Galison, *Einstein's Clock's, Poincare's Maps* (New York, Norton, 2003.
17. Peter Galison and David J. Stump editors, *The Disunity of Science* (Stanford, Stanford Univesity Press, 1996).
18. Peter Galison, *Einstein's Clocks, Poincare's Maps.*
19. Ibid., 251.
20. Alexandre Koyré, *From the Closed World to the Infinite Universe* (Baltimore: Johns Hopkins University Press, 1968).
21. Ibid., 206–272.
22. Ibid., 34, 35.
23. Ibid., 173–175.
24. Etienne Gilson, *The Unity of Philosophical Experience* (New York: Scribner, 1937), 112–18.
25. Koyré, *Closed World*, 6–24.
26. Ibid., 36–38.
27. Weinberg, *The First Three Minutes* (New York: Basic Books, 1993)
28. Koyré, pp. 5, 6.
29. Ibid., 2.
30. Richard Lewontin et. al; *Not in Our Genes* (New York: Pantheon Books, 1984), 277–84.
31. Kant, *Philosophy*, 170.
32. Ibid., 116–117.
33. Ibid., 118.
34. Ibid., 122, 123.
35. Ibid., 147.
36. Ibid., 129–131.

4

Progress by Reduction and the Totalitarian Temptation

Reduction in Full

What may be called the "reductive sensibility" informs modern science, if not as a program for research then as an aspiration. At some point in the future, all scientific explanations, from low-level empirical laws to the most abstract cosmological theories, and in all the variety of fields and subfields of science, will be explicable by one uniform set of scientific laws. Steven Weinberg, Nobel Prize-winning particle physicist, states that for him, ". . . reductionism is not a guideline for research programs but an attitude toward nature itself. It is nothing more or less than the perception that scientific principles are the way they are *because of deeper scientific principles* . . . and that all these principles can be traced to one simple connected set of laws (italics added)." Weinberg goes on to say that particle physics seems to be the most likely approach at "this moment in the history of science." He also admits that "reduction has become a standard Bad Thing in the politics of science."[1]

However diffident Weinberg is as he approaches the topic of reductionism, it is important to note that he is writing in the latter part of the twentieth century. In the middle part of the century, when reduction was not a "standard bad thing," a prominent philosopher of science, Ernest Nagel, explained it as if it were a major component of scientific explanation. Nagel provided both a logical analysis and historical examples to make his case. He also provided a definition: "Reduction . . . is the explanation of a theory or a set of experimental laws established in one area of inquiry, by a theory usually though not invariably formulated for some other domain."[2] Nagel also provided two logical criteria: that the terms of the two scientific fields involved must be explicit and unambiguous, and that the statements of the science being subsumed must analyzable into elementary expressions from which the theory is constructed. Nagel's account is very long and detailed. What is obscured is the fact that, in reduction as it actually takes place, there are two fields of science. One of these is dominant and absorbs the other, a process that is as much a matter of the politics of science as it is of scientific theory. Nagel's primary example is the reduction of thermodynamics to statistical mechanics, or, in effect, of heat theory to atomic theory.[3]

The general idea of reduction may be termed "reductionism" because it is a philosophy and amounts to an ideology, the ideology of science. This notion, following what Weinberg says about reduction being an attitude about "nature itself" and a search for and expectation of "deeper scientific principles," is based on the overall idea that science can and will provide hidden explanations for how things appear at first glance and on the surface. That is, reduction is not only a point about how, as Nagel implies, a scientific theory that explains a set of allied phenomena can be subsumed or absorbed by a more general theory. Rather, reduction in the wider sense of the term, of which Weinberg is aware, is an attitude that affects not just the relations of separate scientific theories, but also how the common perceptions of the prescientific mind-set can and should be explained by scientific principle.

To take the best-known example, there is no question that the common, prescientific perception of mankind immemorially is that the sun makes a daily circuit about the Earth. Now, after the Copernican "revolution," it is the standard mark of an educated person to be aware that it is not the sun that travels but the Earth, which travels relative to the sun. But to accept this means denying what sense perception and human experience indicate, and to perceive the Earth not as a solid, stable mass under our feet but as a globe that is approximately 8,000 miles in radius. At the same time, it is understood that the Earth rotates on its axis once every twenty-four hours, while traveling about the sun in a yearly circle path that is 93 million miles in radius. This double motion, we are told, is the explanation for the alternation of day and night and for the succession of the four seasons at the same time each year. We snicker at the stupidity of the opponents of Galileo, who had to overcome not only theological criticisms and philosophical objections, but commonsense perceptions. But he congratulated those who accepted the Copernican system because in order to accept it, they had to have enough imagination and intellectual will power to accept a doctrine that contradicted their own visual experience on a daily basis.[4] The Copernican revolution is reductive in that it radically transforms the daily, lived experience that is felt, on an existential level, of a static earth as the plane of our daily, "earthly" existence. This includes the transformation of the idea of the beneficent sun, as it travels in its daily circuit overhead and provides heat and light, into a huge ball whirling like a top through outer space. This transformative knowledge is made present to our senses only as an astronomical diagram, an exercise in applied geometry traced on the pages of a textbook.

One French Enlightenment figure went so far as to say that the common impression of our senses, of colors and shapes and of sounds and smell were, simply, *lies*. True reality, as revealed by scientific method, took the form of underlying, *deeper* causes such as atomic activity and molecular combinations. In this manner, reductionism amounts to an alienation of human experience from what science claims is the true reality, a scientific reality that underlies what is the mere surface of physical reality as it is perceived through our sense organs. The author has two memories. The first is of explaining to his immigrant grandmother how one-celled organisms that are invisible to the naked eye cause disease (the germ theory of disease). Grandma, who was an intelligent woman but who lacked a college education, was unimpressed or perhaps didn't understand the concept. In the second memory, as an eighth-grader the author explained to his history teacher, a graduate of Dartmouth, how, over time the radioactive uranium isotope turned into lead. The teacher refused to accept this statement, objecting that lead is used to block radiation from radioactive materials. These examples apply only to physical perception and reality. The contradiction between the prescientific and the scientific mind-set can be extended to the point that science is said to reject the reality of ethical, religious, and artistic insight; it is here that opposition to reductionism is at its fiercest, from William Blake to Martin Heidegger.

Such objections to reductionism avoid the process of reduction as it actually takes place in the history of science. The fact is that the reduction of theories, one into another, does occur, and even though a claim of reduction is often made prior to the actual accomplishment, it takes place frequently enough that its explanatory reality cannot be denied. Further, reduction has an overwhelming importance in the idea of what modern science is and of what it is destined to accomplish. That is, as opposed to our normal and persistent observation that the universe as experienced is made up of such different and varied entities—people, places, and things; the sky above and the earth beneath, in a panoply of colors, textures, and motives—true reality exists in a deeper understanding of a single, all-comprehensive law of nature. In this way, the reductionist project provides the rational basis for the concept of the unity of science.

Reduction and the History of Science

Reduction is not only a doctrine about the scientific explanation of how more complex theories subsume less comprehensive ones, but can be extended into a model of science history and, subsequently, a model of human nature and society. But first, it must be understood that science is not an explanatory monolith, but exists instead in the form of separate scientific fields, each of which works roughly independently of the others. Seen in this manner, modern science readily breaks down into three major fields: the physical or "hard" sciences, biology or the life sciences, and the social sciences. Each of the three major fields contains subfields, such as mechanics and astronomy in physics, and cell biology and classification in biology; there are so many separate types of social science that mentioning their subfields is unnecessary to convey their number and variety.

The reductive model can be understood by means of a relatively simple image, that of three large blocks, made of stone perhaps, sitting on top of one another. The largest of the blocks sits on the bottom and represents the hard sciences, including physics and chemistry. On top of that sits a somewhat smaller but still quite large block that represents biology and the life sciences. The final block, which rests on the former, represents the social sciences. The three-block image represents the notion that physics is the base science on which all the other sciences rest, including even chemistry in the sense that molecular phenomena are ultimately explained by the ability of the atoms that make up molecules to bond together (which, in turn, depends on subatomic processes and entities). Biology is separable from the hard sciences because of the object of its interest and research, namely, living entities. Nonetheless, the processes that underlie the activities of living beings are explained by molecular processes, as the derivation of energy from sunlight depends on the chemical reactions undergone by chlorophyll, etc. Finally, the block that represents all the social sciences sits atop biology, for the individual and social behavior of any kind of living entity depends on its biological form. The biology of ants explains why they live as collectives underground or hidden in tree boles; currently, sociobiology is the scientific subfield intended to form the scientific connections between biology and social behavior.

Between each of the three blocks, in the surfaces between the hard and the life sciences and between the life and the social sciences, there is contact, but not merely that of rough surfaces rubbing against each other.

Rather, between the blocks is a form of cement, so to speak, that provides a connectivity between them in the form of subfields. These subfields can be termed *connector fields*; they include such fields as chemical physics and physical chemistry (within the block of the hard sciences); biochemistry and biophysics, which connect the hard sciences to the life sciences; and neurology and brain science, which connect biology to social science. Currently, the field of genetics is a connector field between chemistry and biology, and evolutionary psychology is a connector field between biology and social science.

In the historical version, reduction is a process that takes place over time. The development of modern science began in the seventeenth century, while the physical sciences, including astronomy and mechanics were established mostly during the eighteenth century. This first phase of historical reduction is represented by the works of geniuses such as Galileo and Newton, both of whom gave mathematical expression and empirical predictability to the motion of observable bodies (and, implicitly, to the action of unobservable atoms). The geometrical curves, or *conic sections*, discovered by the ancient Greeks, including the parabola and the ellipse, were first utilized during this period to predict in the most accurate terms the paths of falling bodies, cannon balls, and planets.

The nineteenth century was the century of the fully reductive life sciences, including the theory of evolution. This occurred at the same time as the discovery of ancient fossils, which came about as a result of carving out canals and mineshafts during the Industrial Revolution. The discovery around the globe of then-unknown life forms arrived with the nineteenth century as a result of sea voyages, such as Captain Cook's and that of the ship *The Beagle*. Darwin was the assigned naturalist researcher on this vessel, as it traveled down the east coast and up the west coast of South America. Classificatory schemata had to be extended to include all the variety of life forms that were then being discovered; evolution was a dynamic interpretation of the extensive schematic "trees" developed by, for example, Agassiz. Genetic science, developed by Mendel, also arose in the nineteenth century, making possible the prospect of reducing living entities to genes and then to molecules and atoms.

In the nineteenth and twentieth centuries, the study of human reality—formerly the province of poets, priests, and philosophers—began on a scientific basis. Anthropology arrived as the result of detailed and comparative studies made of "primitive" cultures discovered in the South Sea Islands, in Australia, in the American plains, etc.[5] Psychology arrived as laboratory experiments were devised to measure such things as human apprehension of color and felt emotions. Disconcertingly, it was discovered that animal behavior is easier to study than human behavior by means of controlled scientific methods. Economics and sociology also arrived during the late-nineteenth and early-twentieth centuries as identifiable fields of modern empirical science, recognized by the founding of professional organizations such as the American Psychological Association in 1892. Finally, the chronological sequencing indicates a sequence in terms of levels of epistemic complexity, from bodies that were observable to the naked eye through the study of astronomy and mechanics (Galileo and Newton), to the complex structures and internal organs of the bodies of living beings (Darwin and Mendel), to human behavior, in which the actions of humans beings had to be captured in a time sequence and in a complex variety of circumstances. In this manner, the chronological sequence moves from single, well-defined bodies such as the moon or a cannon ball studied by astronomy and mechanics, to the complex structures and internal organs of the bodies of living beings, to the infinite variety of human behavior, the understanding of which involves self-understanding as much as observation.

Anti-Reductionist Views

Weinberg concedes that reduction is a "standard bad thing" but does not say why. The answer is twofold: first, it is considered "bad" because of the reaction of critics of a literary, religious, or philosophic bent, and, second, in a more technical way in terms of the internal limits of scientific explanation.

Humanistic Criticism

Reductionism excites both fear and ridicule, but mostly fear, it must be said, because people understandably are afraid of the elimination of those kinds of elements that are felt to distinguish human nature from animal nature, and from the rest of material nature. Whole categories of experience, thought, and belief are ruthlessly eliminated, or so it is felt, from the human essence by science in the reductive mode. Thus, the fear and revulsion is directed not only at the reductive point of view but at modern science as a whole—the scientific enterprise itself, in fact. Such attitudes are not relieved by the remarks of scientists themselves, like Watson, the co-discoverer of DNA, and Skinner, the behavioral psychologist who unapologetically decreed the end of such concepts as freedom of the will and the immortal, immaterial soul. Nor is the fear of reductionism dissolved by the softer words of Edward Wilson, the sociobiologist, when he soothingly described the positive effects of reductionism on art, religion, philosophy, and literature.[6]

In the early arrival of modern science in the seventeenth and eighteenth centuries, William Blake, Mary Wollstonecraft Shelley, and the entire Romantic movement expressed their revulsion and fear quite explicitly, which set a permanent tone of rejection of the reductive point of view and a loathing of the predicted results of the new science. During the twentieth century, the reductive attitude was diligently attacked in C.S. Lewis' novel, *That Hideous Strength*. The title alone gives away Lewis's anger toward the attempt of modern science to co-opt the areas of human imagination formerly expressed in literature, art, and philosophy. Philosopher and aeronautical engineer Wolfgang Smith also rendered a criticism, in his book *Cosmos and Transcendence*, arguing that the mechanical universe of Newton drains color (a mere secondary effect) and meaning from the universe, leaving it flat, uniformly gray, and colorless in scientific fact and emotional affect.[7]

The same general point of criticism was expressed by the cultural critic Bryan Appleyard who, in his book *Understanding the Present*, recommended a policy of existential resistance to the scientific sensibility, but who frankly did not give sound intellectual reasons for doing so.[8] Nonetheless, Appleyard's critique of science in the late twentieth century was perceived by scientific advocates as an attack, and, as such, required an effective response. The most telling and sustained analysis of reductionism came from the cultural critic and writer Theodore Roszak in his book *Where the Wasteland Ends*.[9] Roszak's book is an extended critical analysis of modern life; written in the 1970s, it reflects the attitudes and concerns of that time including a regard for the environment and the hope for a new kind of transcendent politics. Roszak decries the de-mythologization of nature. He favors nature as it was perceived in premodern times, before the gods, ghosts, and occult forces were eliminated by the arrival of Christianity as well as of modern industrialization and science. Roszak's approach is literary rather than philosophic, more critical than positive; among his targets are industrialization and the negative effects of technology, but especially modern science. His view of reductionism comes at the end of a chapter entitled, "Science in *Extremis*; Prospects of an Autopsy," in an appendix "The Reductionist Assault."

Roszak defines reductionism thusly: "I use the term 'reductionism' here broadly to designate that peculiar sensibility which degrades what it studies by depriving its subject of charm, autonomy, dignity, mystery.

As Kathleen Raine puts it, the style of mind which would have us 'see in the pearl nothing but the disease of the oyster.'" Roszak's analysis is unsparingly critical as it portrays the reductive mind-set: "So habitual has this type of knowing become for many scientists and scholars that they seem oblivious to the collective affect of their individual research in depreciating our experience of the world. . . . [Some] obviously take a strange joy in being tough-mindedly irreverent toward what was traditionally held sacred."[10] Roszak extends the analysis by citing the effects of reductionism in a number of research areas, such as "The Automization [sic] of Personality," "Physical Control of the Mind," and "The Nihilism of the New Biology." He quotes, among others, B. F. Skinner, Jacques Monod, and Leon Kass to support these points. While written in the 1970s, it would seem that had he updated his book into the early twenty-first century, Roszak would have had no difficulty in collecting further examples to make his point. For example, brain scientists have recently asserted that they will take the last citadel of human uniqueness once they give a reductive explanation of human consciousness, or use genetic technology, stem cells, and cloning to create new forms of life in the laboratory. These include "chimeras," which combine the features of different species.

Explanatory Criticism

The criticisms of literary figures such as C. S. Lewis and Roszak could be dismissed as the rantings of men who fear science mostly because it is taking over their academic turf. Resistance to reductionism, however, also comes from scientists themselves, that is, those whose fields are the object of reductionist claims from other scientific fields. Weinberg deals with this opposition directly by mentioning controversies he got into with the representatives of two such scientific fields—biology and physics—over his program which reduces all science (and all human knowledge) to particle physics. In Weinberg's version of reductionism, all things depend ultimately on that science that deals with the smallest and apparently most basic elements of physical reality namely subatomic particles; once the atom was split into subordinate parts, they became the ultimate physical basis and symbol of reality. (The fact that subatomic particles exhibit a range of counterintuitive and indeterminate behaviors that contradict major laws of science does not cause Weinberg to reflect that subatomic reality may not provide the most coherent basis for a grand reductive program.)

It is remarkable that Weinberg's subatomic reductionism program was challenged by other physicists—not for theoretical or explanatory reasons, but economic ones. It seems that the kind of subatomic physics that is Weinberg's specialty (he won a Nobel Prize for his research) is so expensive that it eats up enormous amounts of research funding that could otherwise go to other worthy experiments within physics. Weinberg is unapologetic; speaking of what he calls the reductionist world view, he states: "It has to be accepted as it is. Not because we like it, but because that is the way the world works."[11] The primary tool of Weinberg's area of research is the Supercooled Supercollider; the US Congress cancelled the program after spending five billion dollars of federal money on it.[12]

One of Weinberg's challengers was the prominent biologist and historian of biology at Harvard, Ernst Mayr. Weinberg avoids the details of the debate between himself and Mayr, who were both at the time Harvard professors. The issue in question was that Mayr was convinced that on an explanatory level, the laws of biology are not in fact reducible to the laws of physics. The ontological structure seems to be in place, from large organic structures such as internal organs, to biomolecules, to cells, to DNA, to atomic structure, to the internal parts of atoms. Yet biologists have traditionally been adamant that the holistic properties of life processes, or of living organisms are just not explicable in terms of atomic science in any meaningful way. The reductive program in this specific case does not work; Mayr claims that even within biological science, there has been too much emphasis on DNA research, once again involving funding as well as the training of a new generation of researchers being sucked into this new, reductively attractive, but overpraised field.[13] Resistance to the reductive project appears to be strong on the scientific level as well as the cultural and aesthetic.

Hobbes and the Totalitarian Temptation

The link between a mechanical view of human nature and the politics of totalitarianism is more or less obvious to studious observers. The term "totalitarianism" is meant to describe a society that is based upon a dominant characteristic of control by a party, an apparatus, or an elite group whose actions are self-consciously directed by an ideology. Totalitarianism goes beyond mere tyranny in that tyrannies most often depend upon the figure of a single tyrant and thus often dissolve when the tyrant dies or is removed.

Totalitarianism is based more often upon a theory than a personality. Since it is manifested in a party or an elite group made up of many individuals, it can and often does survive the death of the original founder of the regime. Furthermore, as it based upon a theory, totalitarianism can make an appeal to the mass of people whose general opinion must be taken into account; a population in such highly controlled circumstances can only make its opposition known by violent revolt and attempted revolution. The theoretical aspect of a totalitarian regime will consist of an "official" philosophy associated with it, and, indeed, a prominent philosopher to promote and explain it. Examples include Giovanni Gentile (Italian Fascism), Marx (Soviet Communism), and, problematically, Heidegger (Nazism). The theoretical aspect can entice from among the mass of people those idealists who possess intellectual and moral concerns to a greater degree than most others. But common to any philosophy of totalitarianism must be a reductive view of human nature, which is usually expressed in terms of hateful propaganda directed against the putative "enemies" of the state, for instance, capitalists or Jews. In terms of the general population, however, the mass of people in a totalitarian regime are understood not as a collection of individuals with their own dignity, rights, or immortal souls. Instead, they are considered to be part of a vast machine, with no proper reality of their own beyond the measure of the sovereign state; as a result, they are often treated like cattle.

The link between a mechanical view of human nature and totalitarian politics was made explicit early in the development of modern science in the seventeenth century in the writings of Thomas Hobbes. Hobbes is among the most consistent of philosophers, which is apparent in his great work, *Leviathan*. "Leviathan" refers to the state, and it was Hobbes's purpose to create a vision of a state in which internal conflict was not reduced to a minimum but was virtually eliminated altogether. Thus, his state is not realized in terms of a balance of power between the different elements that exist in society such as church, state, guilds, the rich, the poor, nobles, and commoners. Rather, the Hobbesian state is realized by the supreme control of its leader, to whom all citizens have given up their inherent rights. The political aspect of Hobbes's philosophy is so well-known that when readers analyze *Leviathan* they usually limit themselves to its directly political aspects, such as the social contract or the power of the sovereign. Not as often analyzed is the first part of *Leviathan* in which Hobbes propounds a fully mechanical theory of human nature.

The writings of most philosophers are inspired by a general idea, usually a meme that is inherent in the culture of their time. In Hobbes's case, it was the new science of mechanics, which, before Newton, was represented principally by the figure of Galileo. Hobbes must have been enchanted in a measure by the Galilean understanding that the motion of bodies, no matter how complex and mysterious at first glance, could be explained as the application of forces: gravity expressed as the square of time, the flight of cannon balls and planets expressed by the action of two forces—gravity acting downward and forward straight-line motion, which, when combined, could be described in the geometric form of a parabola. Galileo also provided an atomistic philosophy to explain how the universe in full could be understood in this materialistic manner, how all the myriad actions of all the various items that made up the field of human experience could be explained in a mechanical manner as the resultant of separate forces expressed in terms of mathematics and geometry. Hobbes not only studied the new science of Galileo, but made it a point to visit the great man in his old age when he was living under house arrest. Well understood by Hobbes was that Galileo's mechanical view of things eliminated Aristotle's complex causalities, his teleological understanding of the universe, including the actions and parts of animals and of human psychology; his Earth-centered astronomy; and, above all, his doctrine of the soul.

Hobbes relentlessly pursued the expression of his complex vision, beginning, in *Leviathan* with a description of human reality according to the mechanical science of his time. Hobbes begins with a precisely entitled section, "Of Man," which takes up seventeen of the forty-seven chapters that make up the full text. The remaining chapters are on "the Commonwealth," that is, the sovereign state, and contain his concepts of the social contract and the power of the sovereign. In his treatment of human nature, Hobbes applies by a form of analogy the same techniques that Galileo used to explain the motion of physical bodies, which is materialistic in its general method though not made so explicitly. This is important because, at certain points, it becomes apparent that Hobbes is not a consistent materialist nor is his point of view pessimistic, as sometimes alleged. He does, however, reject explanations of human nature by means of philosophic and religious concepts, meaning that Hobbes redefines such concepts as *soul* and *free will* into material terms when necessary. Thus, among the three Christian virtues, he defines hope as "*appetite* with an opinion of attaining," and charity as "*desire* of good to another . . . if to man generally *good nature*" (chapter six). There is no reference to the Christian gospels or the letters of St. Paul, and thus no intimation of the hope of eternal salvation or of charity as a means of obviating deserved punishment for sin.

In his analysis of the state, or commonwealth, Hobbes ignores Cicero, Aquinas, and Grotuis in regard to their development of the doctrine of natural law, which was developed by the ancient Stoics and refined by Christian theologians. Rather, he sets his analytical sites on one observable and generally detectable element of governance—power. He is not interested in justifying the power of a king by divine right but of sovereign powers generally; thus Hobbes found support from both the English kings and Cromwell. Hobbes's logic is unsparing, given that outside of a highly controlled social environment, mankind is in a perpetual state of war of all against all. As he states, ". . . it is manifest that during the time men live without a common Power to keep them all in awe, that they are in that condition called *war*; and such a *war*, as every man against every man." Further, Hobbes notes that "to this war of every man against every man, this is also consequent; that nothing can be Unjust. The notions of Right and Wrong, of Justice and Injustice have there no place. Where there is no common Power, there is no Law; where no law no injustice" (chapter fourteen). Therefore, men must cede their rights to a sovereign power who will keep the peace. Moreover, there is no plausible reason to try to erect a state in the principles of a universal code or natural law inherent in human essence as manifest in social arrangements.

Hobbes's logic specifying the relationship between a mechanical view of human nature and a controlled society established a pattern. In the twentieth century there is, prominently, the example of B.F. Skinner's account, which extends from a concept of human nature defined by behavioral psychology to his vision of a "utopia." Skinner's behavioral version of psychology denudes human nature of its inner essence by eliminating for methodological reasons, consideration of the inner thought and stream of consciousness of human subjects. This reductive view, in turn, makes the concept of a society that is completely controllable through conditioning that is almost entirely positive seem plausible and pleasant. It should be noted, however, that some negative conditioning is occasionally necessary. *Walden Two* is based on a vision of social control that amounts to a kind of "soft" totalitarianism that is enabled not by the use of force or terror, but by an extensive and constant process of social conditioning.[14] Skinner's version of the relation between mechanical human nature and a totally controlled society is reflected in the current policies of national governments, including that of the United States. In pursuing policies of, most notably, the eating habits of the general population and overuse of the environment, governments rely not only on the force of law and regulation but also on encouragements, such as tax relief and government subsidies, as well as outright propaganda, to influence the untutored behavior of its citizens. Such governmental reliance on "nudges" reflects an elitist perception that the general populace is too ignorant to know what is necessary for its own good.

Reduction and Atheism

That modern science exists in opposition to—indeed, seeks to eliminate—religious belief is a general assumption, despite the examples of many prominent scientists who themselves were Christian, including, Galileo, Newton, Pascal, Pasteur, and Faraday, and, in contemporary times, Francis Collins and Owen Gingerich. The presumed "war" between science and religion is probably more a question of social identification than it is an exact intellectual consequence, as a number of philosophers and theologians have erected intellectual systems that contain both the truths about the physical universe discovered by modern science, and the biblical account of God's creative power and providence, among them Teilhard and Polkinghorne. Nonetheless, at its core, modern science is often understood to be atheistic and "anti-God," because an inherent aspect of its project is the demolition of religious belief, particularly Christianity. This understanding reflects a belief in the reductive view of the history of science, which, as advocated by Weinberg, explicitly eliminates any rational basis for religious belief.[15]

The subject of atheism is of current interest due to the arrival of an explicit atheistic social movement in contemporary society; however atheism is a complex subject in all its social and theological dimensions. Even limiting the subject from the point of view of the reductive model will not sufficiently explain atheism in complete terms. But in an attempt to throw light on the subject, I will begin by setting out two categories: *eliminative atheism*, which is the result of explaining everything by means of scientific law, and *replacement atheism*, which reflects an aggressive mentality in which religious belief is directly attacked.

Eliminative Scientific Atheism

Eliminative atheism believes that science will eventually become a closed system of laws and theories that will explain every significant fact about the universe that precludes or eliminates the religious idea. The general religious idea is that there is a spiritual counterpart to sensible reality and that, beyond the veil of the five senses, there is an inner or complementary reality that is somewhat obscure but that in fact provides the essential understanding of what human life is, as well as what ideals mankind ought to live by. The elimination of this spiritual understanding of reality is a large thing to give up on the altar of scientific reason—or so it is often thought. The authority of science is usually brought up at this point in the debate. The objective is to make the case that religious ideas offer nothing that is useful to the scientific understanding of the universe, or to its goal of providing a complete explanation of—and the ability to control—the physical universe.

There is a possible misunderstanding of the famous incident when Laplace was asked by the Emperor Napoleon why there was no mention of God in Laplace's *Analytical Mechanics*, which was an elaborate update of Newton's *Principia*. Laplace famously replied, "Sire, I have no need of that hypothesis," a response that at first glance seems to be a sarcastic dismissal of both God and religious belief, but that has a precise explanatory context. In the *Principia,* Newton's description of the orbits of the planets was based on Kepler's discovery that the planetary orbits followed the path of an ellipse rather than a circle. Despite this correction, subsequent astronomical observation indicated that the orbits were not in fact exactly elliptical, but that occasionally some of the planets went subtly off course from time to time. Newton addressed this imperfection by claiming that God interfered from time to time to correct the planetary orbits in order to bring them back to their intended shape; it is most likely this aspect of Newton's clumsy reliance on God's infinite power that Laplace was referring to.

Laplace's solution was based on the fact, observable by close astronomical observation, that at certain points as the planets move in their orbits they occasionally line up, and do so relatively close together. For example, when the orbits of Jupiter and Mars coincide the two planets are directly in line with one another at their closest point in space. At these points of orbital coincidence, the masses of the two planets pull on one another when, as in this example, the much larger mass of Jupiter pulls Mars somewhat away from its elliptical path. Laplace discovered this phenomenon of so-called perturbation effects, which restored mathematical unity to the solar system.[16]

Laplace is famous for proposing the first explicit ideal of deterministic explanation in science. Wartofsky gives the relevant quote that appears in Laplace's work on probability, commenting that "In such a strict determinism, there would be no unique event which was not included under some law, that is, every so-called unique event would be one of a class of lawful events, determined by the whole state of the system . . . ."[17]

In addition, in such a strict determinism there would be, as it were, no room for divine intervention much less a need for it. Such intervention would amount, in the Laplacean view, to a fatal ripping apart of the fabric of complete scientific explanation, of the network of detailed observation, of inductively arrived-at laws, and of the comprehensive set of well- established theories that make up, in the ideal circumstance envisioned by Laplace and by the reductive model generally, the essence of the scientific project. In this manner, eliminative scientific atheism is an unintended consequence of the reductive model, but atheism in this eliminative form cannot be termed a "project" in that is it does not take the form of a social cause.

Biology and Replacement Atheism

Replacement scientific atheism is observable in the controversies that have been brought up recently regarding the theory of evolution, controversies that have entered the popular culture beyond the bounds of academic discourse. A discernible social movement has come about as the result of its vigorous promotion by several well-known figures, so that at this point atheism may be properly termed a "project." There is a direct link between the recent arrival of replacement atheism and reduction in biological science. One humanistic critic of Darwinism makes a general point about biology and reduction:

> Physicists may dream of a final theory of the universe, but they tend to resist the temptation to extend their ambition to human affairs. Biologists, by contrast, view all of life on a continuum. If so disposed . . . they see nothing in a theory or practice to stop them from moving down along the continuum from the complexity of human life in all its manifestations to the simple beginnings from which complexity was generated. And they may feel no compunction in finding in the simplicity a full explanation for the complexity.[18]

In terms of the three-block model, biology as the middle block can move downward by following the path of biochemistry to include in its purview the basic block of the hard sciences. These days, biology can also move upward, to the social sciences, by means of evolutionary explanations of animal and, necessarily also, of human behavior. In this way, evolutionists can and do claim that they have the best resources not only to contribute to the social sciences, including sociology, but in effect to reduce the whole block of the social sciences to evolutionary biology. Of course, this movement upward into the social sciences has been vigorously rejected; its hubris has been noticed and attacked by sociologists and philosophers alike.[19] Nonetheless, as expressed by such avatars of evolutionary reduction as E.O. Wilson and Daniel Dennett, the range of evolution extends not only to the social sciences but beyond, to enclose those aspects of reality formerly claimed by religious belief. (But while Wilson shows some sympathy with traditional religious belief, Dennett wants it simply eliminated from society![20])

Some of the fire has calmed down in the controversies, but the premise remains effective as a meme in popular culture. Atheism has now "come out of the closet" in our society to become a claimant to the same recognition and deference given to recognized religious denominations, especially the monotheistic ones. Scientific atheism has, in its evolutionary form, become a recognizable social force. Its claim vis-á-vis religion is that, in effect, religion is the first guess at the great riddle of what the essential constitution of the universe is and of what the meaning of human life within it is. As a first guess of barely literate mankind, however, the specific answers provided by religion have irresistibly been replaced, step by small step, by scientific explanations. Thunder represented the anger of Zeus or Thor in ancient times, but now we understand it much more clearly, without the patina of sacred fear, as the effect of electrostatic release among clouds in the sky, and not as an expression of the fearsome anger of some powerful but imaginary god. More importantly, the larger context of religious explanations—with its reliance on many or one god, with nature as a divine artifact that is saturated with emanations, prophetic happenings, and moral lessons—becomes ultimately unnecessary, and is now seen in our advanced age as jejune, a placebo for weak minds.

In this manner, scientific atheism is a social project that claims a special place in current social action as a replacement for religion, toward which it has an aggressive (not to mention hostile) attitude. In the current social climate, however, religious advocates will be energized to rebut the advance of scientific reduction in its newly aggressive form, not so much as an intellectual challenge but as a hostile social movement. In social terms, reductive science has reached its most powerful point so far. This has led, in contrary fashion, not to the proposed unification of all knowledge under the rubric of science, but to one more divisive element in our fractured culture.[21]

Notes

1. Steven Weinberg, *Dreams of a Final Theory* (New York: Pantheon, 1992) 55.
2. Ernest Nagel, *The Structure of Science* (New York: Harcourt Brace, 1961) 338.
3. Ibid. 338–45.

4. Galileo Galilei, *Dialog Concerning the Two Chief World Systems*, trans. S. Drake (Berkeley, CA: University of California Press, 1967), 396–97.
5. See Edward B. Taylor, *Anthropology* published in 1898 for an important example. (New York: Appleton, 1898).
6. Edward O. Wilson, *Consilience* (New York: Vintage, 1998).
7. Wolfgang Smith, *Cosmos and Transcendence* (Lasalle: Sherwood Sugden, 1984). See the chapter "Lost Horizons" for a telling exposition.
8. Bryan Appleyard, *Understanding the Present: Science and the Soul of Modern Man* (New York: Doubleday, 1992).
9. Theodore Roszak, *Where the Wasteland Ends: Politics and Transcendence in Postindustrial Society* (Garden City, NY: Anchor Books, 1972).
10. Both quotes, p. 242.
11. Weinberg, *Dreams*, 56.
12. Ibid, 276–90.
13. Ibid, 259–60.
14. B. F. Skinner, *Walden Two*.
15. Weinberg, *Dreams*, 56.
16. http://www.encyclopediaofmath.org/index.php/Perturbation_theory.
17. Marx W. Wartofsky, *Conceptual Foundations of Scientific Thought* (New York: MacMillan, 1968), 298.
18. Eugene Goodheart, *Darwinian Misadventures in the Humanities* (New Brunswick, NJ: Transaction Publishers, 2007), 89.
19. Criticisms; see John Caiazza, "Political Dilemmas of Social Biology," *Political Science Reviewer*, Vol. xxix, 2005.
20. E.O. Wilson, *Consilience,* 260–90; Daniel Dennett, *Darwin's Dangerous Idea* (New York: Simon and Schuster, 1995), 515–18.
21. John Caiazza, *Disunity of American Culture* (New Brunswick, NJ: Transaction, 2013).

5

Historicism, Relativism, and the Open Society

The prior three chapters have presented variations and nuances of the Whig idea of scientific progress and their social implications. But toward the end of the twentieth century, a new concept of the career of modern science developed, one which exhibited profound doubt about the reliability of science as the truth teller of modern civilization, and that, in turn, implied a profound doubt about the general concept of scientific progress. This view of modern science also has significant implications for the concept of social values, reflected in its doubt about the reliability of ideologies, both totalitarian and democratic. This "historicist" view is presented in this chapter.

From the Philosophy of Science to the History of Science

In the middle of the last century, the philosophy of science developed into an identifiable and separate field of philosophy. It had developed its own journals (for example, *The British Journal of the Philosophy of Science*) and its own institutes and programs in established universities, including Indiana University and Boston University. The development of a separate field reflected the acknowledged importance and influence of modern empirical science, which had become especially noteworthy in the twentieth century due to the advancement of new comprehensive theories. The significance of the field also stemmed from the enormous technological developments that have affected and transformed all areas of human life from agriculture to electronic communication to birth control. What mainly intrigued philosophers of science as the field emerged was not the social effects of technological developments or the emergence of such theories as neo-Darwinism and Relativity, but issues related to the nature of scientific explanation and associated questions of scientific logic and proper methodology. Here the inspiration came largely from the positivism of the Vienna Circle and philosophers such as Russell, Quine, and Carnap, whose philosophical approach relied heavily on the explication of the empirical philosophical implications expressed in terms of formal logical structures.[1]

A representative example of such a philosophy of science is Ernest Nagel's *The Structure of Science*. Nagel, who was a prominent philosopher, wrote in the preface to this large and imposing work (just over 600 pages of text) that, "its scope is controlled by the objective of analyzing the logic of scientific inquiry and the logical structure of its intellectual products. It is primarily an examination of logical patterns exhibited in the organization of scientific knowledge as well as of the logical methods whose use . . . is the most enduring feature of modern science."[2] This formal approach, which emphasized scientific methodology, was challenged in the 1960s by the writings of several philosophers of science who imported into the field an approach based on the historical and social aspects of modern science. During the 1970s, the positivistic influence of the Vienna Circle was being supplanted by the historicist influence of several authors, including Paul Feyerabend, Michael Polanyi, Norwood Russell Hanson, and Karl Popper but especially Imre Lakatos and Thomas Kuhn.

Polanyi emphasized the degree to which scientific knowledge was personal knowledge, tacitly known, and an acquired skill. Feyerabend attacked the ideal of scientific rationalism, forcefully denying that the events, procedures, and results that constitute the sciences have a common structure. Hanson discovered patterns of discovery and invented the term "theory-laden observation terms" to emphasis the point that scientific observations were not neutral descriptions of experimentally discovered "facts" but also included a preexisting set of expectations.

> To say that Tycho and Kepler, Simplicius and Galileo, Priestly and Lavoisier . . . Heisenberg and Bohm all make the same observations but use them differently is too easy. It does not explain controversy in research science. Were there no sense in which they were different observations they could not be used differently. This may perplex some: that researchers sometime do not appreciate data in the same way is a serious matter. . . . There is a sense then in which seeing is a theory laden undertaking.[3]

Overall, the new, historicist emphasis on the personal and the cultural implied that scientific reason could no longer plausibly be thought to fit Enlightenment ideals, a belief that has had a direct impact.

The transition of the academic study of modern science from a positivist to a historicist venue may be clearer as a description of what was left behind than what the study of modern science has become. That is, while the positivism of the Vienna Circle is understood well enough to provide readers with a good idea of what it means, the terms "historicism" most likely does not have the same degree of clarity, especially as it is being applied here to the study of science. In order to make it somewhat clearer, in the next several sections the term "historicism" is analyzed with respect to three elements, each of which is discussed with reference to the history of modern science. These are: the perception of *deep patterns* in the overall history of science, *global wholes* or world-views as the means of understanding the scientific mind-set, and *Idealist epistemology*.

Deep Patterns: Vico

Bacon, the major prophet of modern science, proclaimed the ongoing increase of scientific knowledge, and described it as the expectation that science as a social enterprise would grow by the accumulation of empirical facts and inductively derived theories. The Whig concept of progress and its immediate variations, the Enlightenment and Reductive concepts, share the expectation of the increasing power of scientific knowledge, which will culminate in a "Theory of Everything "currently pursued by contemporary physics. At the heart of the common understanding of modern science is the idea of progressivity, that science is always making inroads into hitherto unknown and seemingly inexplicable areas of reality. What is today unknown is not unknowable, and there are no impregnable defenses against the increase of scientific knowledge into the formerly reserved areas of mind, human perception, the soul, and God (or at least the origins and causes of religious belief). Current advocates of a scientific point of view, such as, Weinberg and Dennett, assert that all the seemingly mysterious and unique aspects of human existence will be comprehended in time by the scientific method. The whole discussion regarding such strange entities as *strings*, *membranes*, extra dimensions, and multiple universes is an expression of the sense among many physicists who work in the most abstract of scientific realms that this ideal of culmination is on the verge of becoming a reality. (But does not the appeal of such ultimate explanations lie in their very weirdness? It is as if the fairy-tale strangeness of these abstract entities is validation of their truth.)

In the century following Bacon, Vico put forth an account of human knowledge and experience based on a rejection of the principles of Enlightenment rationalism and of science. Vico is notable in this context for two major reasons. First, in direct resistance to the Cartesian ideal of knowledge based on clear and distinct ideas and deductively derived principles, he claimed that truest knowledge was that which human beings created for themselves. This, Vico asserted, was found not in modern science but in literature, poetry, a study of laws, and the customs of people ("philology"). In contemporary terms, Vico's ideas meant that a study of culture and not of empirical science provides the best means of understanding humanity.

The other aspect of Vico's thought was his complex account of the deep pattern of human history; Vico discerned three stages of the development of civilizations, each of which has its characteristic representation in the areas of the concept of nature, custom, jurisprudence, reason, etc.[4] The three-stage pattern is universal in human history and takes place throughout all the various civilizations of mankind. Such a thesis, however, is not new with Vico. For his source, he references the Egyptians and the Roman poets, who wrote of a past golden age, followed by a silver, an iron, and, finally, a leaden age—an account of human civilization in reverse order, as it were. But Vico had an optimistic view of the course of human civilization, writing as he did at a time of great advancement in human knowledge in the areas of medicine, empirical science, and the arts (such as the discovery of perspective). The new thought that Vico provided about universal history is useful to consider in the present context of examining science history. Vico's account was not merely progressive, as it worked through stages (as did Comte's), but also included an exploration of the decline of civilizations.

What is most distinctive about Vico's account of universal history is his assertion that, after they arise from barbarism and then undergo the general pattern of development, civilizations lapse back into a new form of barbarism. But this is a new, refined form of barbarism. According to Vico, from such new and different forms of barbarism, new civilizations arise that then follow the aforementioned three-part pattern of development. Vico's pattern of civilizational development is a spiral that ascends upward, combining the path of a circle with that of an upward rising line. This is Vico's famous theory of the *ricoursi,* more complex certainly than the straight- line model of ascent that is at the basis of the Whig account of the history of science. Vico's recursive model of history, in which certain eras repeat themselves in a transformed fashion, might not seem to have much application or consequence for the history of science; yet, examples that fit his recursive theory can be found, for instance, the history of theories of light.

The history of theories of light has been referred to in the discussion of spectrography in chapter two. It is well-known, and only needs to be mentioned in brief, that Huygens's wave of theory of light was dismissed with the success of Newton's particle theory, and that one hundred years later Fresnel and Young revived the wave theory. But, as in the case of Vico's barbarisms, the new version of wave theory was distinctively different from its earlier manifestation. For as the wave theory was taken up in the nineteenth century, it was subsumed into Faraday's concepts and Maxwell's field equations. Light waves were transformed into electromagnetic waves along a spectrum, and the wave activity of light connected to the forces of electricity and magnetism. But the development of light theory doesn't end with the direct reascension of wave over particle theory. The final stage is indicated by Einstein's account of the photoelectric effect. Einstein explained this effect by means of the theory of light quanta in the form of photons, which, in the subsequent development of quantum mechanics, was describable by either wave mechanics or matrix mathematics, that is, as either the action of waves (of probability) or particles (packets of energy).

The conflict, or competition, between the particle and wave theories of light did not result in the dominance of one over the other because by the late-nineteenth century detailed experimental evidence regarding light spectra was linked to atomic theory and quantum physics. Newton's research into the light spectrum had been only the first step in further productive research in which different spectra were produced by burning certain metals and chemicals, and in which exact patterns of spectral lines could be observed and used to identify the different chemical elements. Then a connection was made between the spectra of each separate element and the number of electrons that surrounded its atomic nucleus. The resolution of the debate rests in the special peculiarities of quantum physics, in particular, wave-particle duality in which light can be seen either as particles, that is, photons, or as wave emanations, depending on the physical context.

Rhythms are discernible in the history of science in terms of its theories in larger and somewhat less definable turns. The light-theory example operates on identifiable historical evidence but indicates that such patterns of opposition are resolved in the history of science by neither domination nor rejection. Instead, after consideration of further and detailed observation under the rubric of new theories such as quanta and relativity, the patterns of opposition are explained as separate but related facets of the same phenomena. In accord with Vico's insight, patterns in science history that had been discerned by Kuhn and by Lakatos, among others, include philosophical and cultural elements as well as strictly empirical or theoretical ones.

This brief account indicates not merely that the two theories in competition show a pattern in which they ascend and descend in relation to each other. It also illustrates that in the end they are, roughly speaking, combined in the complementary manner of transferable but different mathematical expression in quantum mechanics, and later encapsulated into quantum electrodynamics (Feynman). Could such a recursive pattern occur in other areas of science history, and possibly in the history of modern science as a whole? And, if so, what is the significance? These questions will come up again in subsequent chapters.

Global Wholes: Kuhn

So much has been written about Kuhn that it seems impossible to encapsulate his reputation, thought, and influence into a coherent pattern. A general assessment will not be attempted here. Instead, what is offered is a perhaps unique take on Kuhn's thinking—not as sociology, radical critique of big science, or as weak philosophic analysis as seen from the point of view of empiricist philosophy. Rather, what is offered is a presentation of Kuhn's thought as an unintended form of historicist philosophy. In other words, there is more to be learned about Kuhn's thought from the tradition of what is termed the philosophy of history (for example, Vico, Hegel, Dilthey, Collingwood, and others) than from empiricist logic (for instance, Carnap, Quine, Russell, among others).

The take from the tradition of the philosophy of history is most useful in unpacking the sense of Kuhn's most evocative and original concept, that of *paradigms*. Without question, this concept is the most characteristic and controversial element of Kuhn's philosophy. The concept of paradigms was used to expand scientific explanation to include (beyond the putative empirical truth of a theory) the social aspects of human belief systems in general. This was especially true as these were influenced by the reputation and intellectual power of basic, foundational texts. The expansion went on to include the technical manner that an identifiable set of researchers used to perform experiments, and the mathematical forms they used to erect physical laws. Here Kuhn explains why he decided on paradigm analysis: "By choosing it, I mean to suggest that some accepted examples of actual scientific practice – examples which include law, theory, application, and instrumentation together – provide models from which spring particular coherent traditions of scientific research."[5] It is notable that despite his definition and subsequent 200 pages of development that the concept of paradigms resists satisfactory description; one critic teased out ten different meanings of the term![6] What is offered here is an unpacking of the concept in terms of *global wholes*, the necessary term of art of the tradition of the philosophy of history.

The historicist methodology that Kuhn pursued rests on descriptions of the course of successive paradigms in terms of their global aspect. Isaiah Berlin wrote that, for Vico in particular and for historicist analysis generally, ". . . there is a pervasive pattern which characterizes all the activities of any given society: a common style reflected in the thought, the arts, the social institutions, the language, the ways of life and action, of an entire society."[7] Likewise, paradigms are "global wholes" since they are not only dominant ideas on an intellectual level. In addition, as described at length by Kuhn, paradigms involve social identity, a deep sense of personal commitment, an experimental methodology, a means of understanding (in effect, their own system of rationality), and a tool kit for solving problems; change from one Paradigm (global whole) to another is a matter of personal conversion, social revolution, or both. In general, the methodology of a philosophical history—historicism—exhibits a distinct and well-established pattern, for analysis in terms of global wholes is apparent in the writings of Vico, of Hegel, and of Collingwood. While philosophical history is done in terms of deep patterns over time, it is also carried out by describing the dominating element that constitute these patterns as holistic entities, that is, as global wholes that dominate their respective cultures. Thus, Hegel's description of the dominant global whole of his time—the "weltgeist," or world spirit—included, besides "mind," law, morality, art, revealed religion, and philosophy.[8] The use of a global or holistic means of historical description is obviously apparent in the scientific paradigms of Kuhn. However, he limits their application to the history of science, an approach that is likely to be more philosophically effective than grandiose attempts at analyzing human history in all of its aspects, historical patterns, and global wholes.

The concentration on global wholes has wide epistemological implications that are in effect, anti-Cartesian. That is, the human mind does not approach human things in the manner of a laboratory researcher who studies the results of an experiment, but rather by a form of self-analysis. Berlin explains, ". . . men's knowledge of the external world which we can observe, describe, classify, reflect upon, and of which we can record the regularities in time and space, differs in principle from their knowledge of the world that they themselves create, and which obeys rules that they themselves have imposed on their own creations."[9] This implies a sharp distinction between the kind of knowledge that is available through, for example, the hard sciences of physics, mechanics, astronomy, and chemistry versus that attained by psychology, literature, poetry, and philosophy. Kuhn's philosophy of science tends strongly to eliminate this distinction, so that all knowledge, including scientific, is attained via the medium of the paradigms.

Historicist Idealism and Its Critics: Scheffler

The historicist model is not the construction of only one thinker (Kuhn), but rather represents an intellectual and cultural movement in the manner in which science is viewed, that is, from a positivist to a historicist turn. There is no avoiding that the historicist turn has a direct impact on the view of scientific truth; as has already been said, historicist philosophy implies an Idealist form of epistemology. In the most extreme terms, truth is not definable as the relation of opinion to reality, since reality itself is presumed in the historicist mind-set to be a rational construction. Therefore, there is no independent basis for scientific truth outside of theoretical terms and their relations, which modern science sets out as its ultimate explanations of areas of research. The historicist conception of truth rests on the coherence theory that statements about the universe must be consistent with other such related statements, for what defines their truth is the degree of their consistency or necessity as logically derived consequences that are based on a main theory.

The historicist conception of truth directly confronts the "standard" view of philosophy of science. Up to that point in time, this view was based almost exclusively on the idea that empirical evidence drawn from the basic activity of modern science—controlled observation and exact measurement observed repeatedly—leads to the discovery of a general law of nature. Resistance from the "old guard" came swiftly, representative of which was the publication of *Science and Subjectivity*,[10] a short but well-written book by Israel Scheffler, who was a professor of education and philosophy at Harvard University. The book begins with a chapter, "Objectivity Under Attack," that expresses Scheffler's exact idea of what is at stake. The chapters that follow focus on the topics of scientific observation, and the meaning of theories and theory change, while the concluding chapter examines "the epistemology of objectivity." In the first chapter, on which I will concentrate here, Scheffler explains the importance of the issue, stating that ". . . objectivity is the end, as well as the beginning of wisdom."[11] He proceeds to pointedly examine the three authors—namely, Hanson, Kuhn, and Polanyi—whom he presumably sees as most representative of the anti-objectivity movement as he understands the historicist trend then under way.

Of Hanson's idea that observation terms in science are "theory laden," Scheffler states that

> . . . if seeing is indeed theory laden in the sense described [Hanson's sense], then proponents of two different theories cannot observe the same things in an effort to resolve their differences; they share no neutral observations of deciding between them. To judge one theory as superior to the other by appeal to observation is always doomed, therefore, to beg the very question at issue.

Of Kuhn's idea of scientific change through revolutions that overturn paradigms, Scheffler states that such an account implies that "[t]o understand another's apparently observational or experimental references, we must first enter into his theoretical thought-world." Scheffler concludes that "the breakdown of observational community and of the community of meaning, and the consequent rejection of cumulativeness, seem to remove all sense from the notion of a rational progression of scientific viewpoints from age to age." To this harsh judgment of Kuhn's paradigms, Scheffler adds another against Polanyi's concept of tacit knowledge, which concludes his uncompromising attack on the historicist turn: "The general conclusion to which we appear to be driven is that adoption of a new scientific theory is an intuitive or mystical affair, a matter for psychological description primarily, rather than for logical or methodological codification."[12]

Thus far, Scheffler has citied the testimony of three hostile witnesses like a prosecuting attorney; he presents his final charge against the enemies of scientific objectivity in the next-to-last paragraph. If their view is to be taken seriously, Scheffler argues, then "independent and public controls are no more, communication has failed, the common universe of things is a delusion, reality itself is made by the scientist rather than discovered by him. . . . I cannot myself believe that this bleak picture, representing an extravagant idealism, is true. In fact, it seems to me a *reductio ad absurdum* of the reasonings from which it flows."[13] But is the case really this solid, and, we may ask, is there a description of the implications and effects of the historicist view that is less accusatory and more sympathetic?

In order to answer this general question, we must recognize that there are really two issues here. First, the issue of what reality must look like or be in order for the historicist account to be the best available account of science. Scheffler is not wrong in claiming that the historicist view of modern science is a form of philosophic idealism, but it is not, as he states, an "extravagant" idealism; his point applies to any and all historicist descriptions of any kind of history when brought to their metaphysical conclusion. Idealism in history implies an idealistic metaphysics, which was most clearly, if not extravagantly, represented by Hegel's philosophy of mind in which ultimate reality is described in terms of an idea, that is, *weltgeist*. Hanson, Kuhn, and Polyani never get this far in their thinking, and leave the matter unsettled.

Beyond this metaphysical and epistemological issue is a second concern. Namely, is Scheffler correct when he states that Kuhn's paradigmatic and revolutionary account of scientific history implies the breakdown of scientific meaning and "the consequent rejection of cumulativeness" that seems to remove "all sense from the notion of rational progression of scientific viewpoints from age to age"? This is the more relevant issue of the two in examining a historicist model of scientific progress; if correct, Scheffler's criticism implies that the paradigm model undercuts any attempt to erect a rational history of science. In effect, the paradigm history of science is a series of paradigms that go nowhere. The idea of an end of scientific history in the form of a final theory is neglected and effectively denied, and so the history of science will seem like a history of European warfare, from Caesar to Charlemagne to Napoleon to Hitler, each chapter interesting in itself, perhaps, but arriving nowhere in the end.

Meaning and Science History: Popper

Despite Scheffler's criticism and that of others, the group of philosophers we are terming "historicist" did not develop their approach out of a misplaced desire to insert a neglected philosophical tradition—Idealism—into our understanding of modern science. Rather, Hanson, Kuhn, Polanyi, and others came to their viewpoints by a study of science as it is actually done rather than according to the rubric laid down by Bacon at the start of the scientific revolution. That rubric amounted to a tradition that, like many old traditions, had been outworn by circumstances, namely the relativity and quantum revolutions in physics that took place at the beginning of the twentieth century. Philosophers who attempted to understand science could no longer assume that the inductive process of discovery, which occurred over a long period and was based on the research activities of groups of like-minded naturalists and scientists, would in a gradual and effective manner eventuate in a grand theory of everything. (Popper called this account "Inductivism," which will be discussed later in this chapter.) Rather, the process of science history was *not* a more or less smooth one, proceeding as it did in fits and starts, with the inevitable end of ultimate scientific clarity in view. Rather, Kuhn and the historicists insisted, the process of science history was subject to great and unexpected changes that were properly understood as "revolutions."

The signal event that precipitated the change from a positivistic philosophy of science to a historicist point of view was the arrival of the theory of relativity. Relativity was not a development of the mechanical theory of Newton that had virtually directed the theoretical developments of physics for almost 200 years. Rather, it was a radical displacement of it that, as Kuhn stated and that appeared in a passage quoted by Scheffler, meant that the terms "mass" and "gravity" had significantly different meanings in the two competing paradigmatic theories of Newtonian mechanism and Einsteinian relativity. Hanson was being historically accurate when he stated that Tycho and Kepler "see" different things when they observe the same sunrise; certainly followers of Ptolemy and Copernicus did. Nor was Polanyi incorrect when he observed that theory change amounted to a form of "conversion" in the sense that the change of theory implied a new form of tacit knowledge and of how experience itself is organized; certainly that was the case for Galileo when he moved from the Ptolemaic to the Copernican system. The great historian of the nineteenth century, the German scholar von Ranke, said that the aim of history was to state what had happened "*als virtlich war*," as it really was. What drove the new emphasis on the history of science among philosophers and scholars was their discovery of how science is done, as it really was, and is.

But what was to be the overall effect of replacing a positivist understanding of science with the historicist point of view? First, there are epistemological consequences, for a new concept of how scientists actually observe experimental results had to be devised, one that allowed for theoretical errors to evolve *even when the process of inductive generation of general laws and theories had been done correctly*. Second, a new account of the general lines of the history of science had to be devised to account for the reality of fundamental theory change, one that was not automatically progressive as in the Whig theory of scientific progress. Third, a new concept of the relationship of modern science to the Enlightenment and to Western civilization was implied, for science could no longer be assumed to be the engine of social progress and democracy. The first to detect these consequences was the philosopher Karl Popper, from the 1930s to the 1950s.

Popper's philosophical position was as an outlier of the Vienna Circle; Popper circled around it, taking in the triumphant beliefs in the progress of scientific knowledge as promoted by Schlick, the Circle's chief organizer, and its major figures, such as Carnap. Having digested the significance of Einstein's relativistic revolution, however, Popper worked diligently and with great creativity to develop an account of scientific knowledge that would do two things at once. First of all, account for the fact of a revolutionary theory change and of its future possibility in modern science, and secondly not slide off into a form of "mysticism" or irrationalism that might justify speculative philosophy and religious belief. That is, up to that time, the validity of modern science as the premier form of knowledge and the most reliable way to understand the universe was based on the surety and content of its major theories, particularly mechanism and evolution. But with the Einsteinian revolution, the validity of these major theories could no longer provide the source of rational surety of science. But Popper was unwilling to give up the rationalism promised by the advance of science, and so his response was to replace the theoretical content with the method of modern science. Hitherto, Popper argued in effect, science's validity was not to be found in presumably final pronouncements of major theories, that is, in science's truth-telling capacity, but in its reliance on a specific type of scientific method. But to accomplish this, Popper needed to reimagine what constituted the method of modern science.

Much of the work being done by members of the Vienna Circle at that time was in the area of *confirmation*, the standards, methods, and protocols that are required to prove scientific statements, laws, and theories true. Popper argued in his important book, *The Logic of Scientific Discovery*, that the real work of science and the reason for its progress had been misunderstood; it was not that science is devoted to confirmation of its statements but to their *falsification*. Popper's falsification thesis proposed that scientific progress is made when scientists approach an area of research, and define it and its associated problems by means of statements that can be proven false. This, in turn, implies that such statements are very clearly constructed.[14]

The implication for defining science's path to progress was left unclear in Popper's account, which may well have been deliberate. Popper refused to countenance any idea of patterns of progress in the history of science or in any other area of human reality, such as economics or politics. He not only discounted but attacked "historicism," root and branch, namely, global wholes and deep patterns. Popper expressed his hostility toward historicism particularly in a later, much shorter book, *The Poverty of Historicism*.[15] Thus, Popper, with his aversion to anything smacking of irrationalism, ended historically in the same place as Kuhn and the historicists themselves. Although he had avoided what he perceived later on as the irrationalism of the historicist approach, he, just as much as they, could not provide a clear account of the path of the history of science. Unlike the historicists, however, his inability to discern such a pattern was the deliberate and understood consequence of his philosophy of science, not a matter of neglect. Nonetheless, they had ended up in the same place, since historicists and Popperians alike are akin to strangers set in the middle of a large, bustling city without a map or a train schedule.

Historicism and Popper's Contentless "Open Society"

Popper stands out among philosophers of science in the twentieth century as the only major thinker who detected the social and political significance of the historicist turn; no other philosopher of that century was able to make such a connection, By now, of course, the perception that there is such a connection—between scientific epistemology and political ideals—is well-understood, if not commonplace. Popper was forced to leave his home city of Vienna with the rise of the Nazis. (Popper's parents were born Jewish but had resigned their Jewish heritage to become Lutherans, the better to fit in with the dominant religion. Popper himself was irreligious, a fact that would not have availed him given that the basis of Nazi racial policy was based on ethnic inheritance and not religious identity exclusively.) Like many others in that situation, Popper was desperate to get to England; he was assisted in this effort by the economic theorist Friedrich Hayek, who was established there.

During the war years Popper had a teaching position in New Zealand during which he wrote his most influential book, *The Open Society and Its Enemies*.[16] In the two volumes that were published under that title, the first was mainly on Plato; the second, on several other thinkers including Hegel and Aristotle; however Popper's main criticism in the second volume was aimed at Marx, an attack that has provoked much response and criticism.[17] Popper's destructive analysis of Marxist theory and practice was an act of serious courage, for in those days premature anti-Communists were excluded from academic positions and polite society, which meant a loss of influence and employment for such early anti-Communists as Orwell, Malraux, and Chambers. What Popper had come to understand through his personal experience, no doubt prior to his developing a philosophic account, was that despite their contradictory political theories, differing cultural traditions (one cosmopolitan and enlightened, the other nationalistic and retrogressive), and bloody political opposition that reached warlike levels in the city streets of Eastern Europe, both Nazism and Communism were totalitarian in nature. Both systems denied liberty to the individual and suppressed political opposition by the most cynical and brutal means, in the service of an overriding ideology that could not withstand sustained intellectual analysis. Above all, when in full effect, both systems took over entire nations and killed millions of people in the effort to construct a new vision of society. This insight is reflected in Popper's dedication to *The Poverty of Historicism*: "In memory of the countless men and women of all creeds or nations or races who fell victims to the fascist and communist belief in Inexorable Laws of Historical Destiny."[18]

The underlying source of Popper's thought, which is consistent to a remarkable degree as it extends from scientific to political philosophy, is his understanding of historicism. But Popper relies on an odd definition of historicism (which, in part, agrees with that of his mentor, Hayek): "I mean by 'historicism' an approach to the social sciences which assumed that *historical prediction* is their principle aim, and which assumes that this aim is attainable by discovering the 'rhythms' or the 'patterns', the 'laws' or the 'trends' that underlie the evolution of history."[19] He applies this criticism to the social sciences, in which it is generally believed that the issues of explanation and prediction are more complex and troubled than in the hard sciences. Thus, Popper's account would seem to be raising a mainly methodological issue and, frankly, not a very significant one. Moreover, the criticism is not entirely accurate, the aim of historicism is to achieve better understanding, which might lead to the ability or desire of the historicist historian to attempt prediction, without its being the principle aim. For Popper however, and for Hayek, historicism is allied with totalitarian social ideology, and in the non-falsifiability of the basic statements of such ideologies lies the necessity of their adherents to impose their utopian vision by the harshest means on their suffering populations.

Popper's critical analysis seems to apply more easily to Marxist political philosophy and Communist applications of it than to Fascist ideology, which tends to discard a rationalistic approach. The appeal of Marxist socialism to intelligent and well-meaning people stems from its deeply constructed historicist theory—of the evolution of society by means of production, of the necessary development of classes and of class warfare, and, it needs to be stated, of a moral concern for the downtrodden of society. But however theoretically compelling and well-intentioned, in practice Marxist ideology demonstrates an absolute unwillingness to accommodate any opposition to its manufacturing of a "heaven on earth," resulting in the most brutal and cynical application of forces including gulags and forced starvation.

Despite the cogency of his argumentation, Popper in some sense seems to have lost the point of his own analysis, for in his view it is not the content of scientific theories that manifest truth and it is not a totalitarian political ideology that guarantees social amity. In Popper's thought, scientific epistemology and political philosophy are intimately linked, but the linkage automatically guarantees failure of the intellect's ability to know whether any general statements are true. In the final analysis, no one can be sure of the definitive truth of any scientific theory, no more than they can be sure of the truth of a political ideology. The final effect of Popper's overall philosophic approach is somewhat skeptical, like Hume's, and his own political view reflects his scientific epistemology. As much as the content of scientific theories is neglected or set aside, so also is any content of a true or adequate political ideology or system left unmentioned or not described. Popper advances no theory of what a good society would look like (unlike Rawls or Nozick), nor does he render an account of human nature, or of nature generally, as it is studied by modern science (unlike Peirce and Whitehead).

There is a vacuousness in Popper's philosophy, for the fact is that scientists do care whether the theories to which they are committed are true or not, that is, are confirmed; they are not really in the business of putting forth falsifiable statements but rather true statements that they are aware must be tested by further research. So, also, even as it results in a claim against the power of the totalitarian state, Popper's political theory lacks content since it does not specify in any detail what attitudes or ideology ought to underlie a democratic state. Popper aims at defending and promoting his vision of the open society but does not specify what it looks like. Interestingly, and by contrast, Popper's friend Hayek, whose influence can be detected in Popper's political thought, did specify the content of a democratic society.[20]

The Open Society, Right and Left

Popper's concept of the "open society" may be contentless, but it is plainly persuasive to many people because it is responsive to what is perhaps the primary general assumption of contemporary democratic politics. That is, underlying the idea of what constitutes democratic society today is the sense that any categorical assertion of any doctrine or value that is applied to society or individuals is immediately suspect. Popper's approach fails to state the content of what constitutes liberty, good governance, human nature, or beneficent social behavior. An epistemological approach that limits itself to criticism and that neglects positive statements leads to the conclusion that there can be no intellectually justified or morally correct description of the open society, which leaves it open to wide interpretation.

Oddly, this lack of content of what constitutes a democratic society is preferable to many thinking people in the West, for it is as if the mere fact of an assertion of a categorical statement is enough to provoke resistance without regard to the actual content of the statement. Such assertiveness on the part of adherents is thought to constitute the first step toward a totalitarian and unfree society, or so it is felt. It is as if what Hitler said about race is not wrong because, in fact, there is no such thing as a "Nordic race," much less that it is superior to any other races and that such belief provides a motive for the elimination of other races. Or as if what Lenin said about class warfare is not wrong because it is an evil thing that wrecks societies and causes illimitable hatred and civil war. Rather, it is the fact that Hitler and Lenin said anything at all, which they presumed to be true assertions that applied to social organization, that is the true fault of their ideologies.

And so we are encouraged in contemporary Western society not to take any political philosophy too seriously, and to avoid making statements in public about, for example, the American tradition of freedom, the British Constitution, Christian morality, etc. Despite the lack of specified content in the open society, however, serious consideration must still be given to what moral principles will underlie the policies and the ethos of democratic society. As a result, political thinkers on both the right and the left have supplied the emptiness of the open society with their own belief systems. But formulating them in the context of an overall social sentiment that decries absolute statements will obviously inhibit or distort such belief systems. This constriction of social thought shows itself in the manner in which the traditional breakdown of political thinking in the modern West is expressed, that is, in terms of "right" and "left."

A drift toward the right is apparent in Popper's thinking. Given its complexity and nuance, Popper's political philosophy cannot easily be described in terms that are familiar; it is not "liberal" in the contemporary, sense since government programs of the sort favored by the left are based on general ideas that Popper would find non-falsifiable, namely, metaphysical and beyond rational proof. Nor is Popper's political philosophy "conservative" in the sense that conservatism implies a reverence for tradition and received patterns of culture; as a devotee of scientific rationality, however redefined, Popper would not countenance such an attitude. But Popper's political philosophy can be accurately described as "libertarian," a somewhat new political attitude that is gaining in popularity in England and America. Libertarian thought focuses on freedom for the individual and maintains a sharply critical stance toward political power and social attitudes that seek to control individual behavior. Because of this, Popper was seen as a premier intellectual supporter of the Thatcher government in Great Britain and was celebrated as such.[21]

Regarding capitalism, Popper is not as much of an advocate of the free market as Hayek was. But Popper's very lack of an overriding political philosophy means that Popper, in effect, accepts it as the consequence of social freedom. That is, while Popper proposes no actual argument on behalf of free-market economics, he apparently accepts it as one of the elements of a free, in other words, open society. By implication, Popper would argue that constraints on economic activity, such as those advocated by Marx and other socialists, are based on non-falsifiable and historicist concepts that would inevitably lead to a closed society. The actual effects of such policies, as seen in the Soviet Union, serve as a warning. But, to reiterate, Popper is reluctant to offer an argument on behalf of the free market as such, contrary to contemporary thinkers such as Hayek, Nozick, Friedman, and others.

Popper's libertarian view of the open society is not the only one possible, for a left- wing version is available that is expressed in the activities of George Soros, the financier and social activist, and in the writings of Richard Rorty. In contrast to Popper, Rorty has a positive view of historicism that resides precisely in what to many critics is its greatest fault, namely, its tendency toward relativism. While for Popper, historicism becomes a repository of justifications for totalitarianism, for Rorty historicism works to undermine all claims to certainty of knowledge, whether in the scientific or political realm. In this manner of extending the analysis from science to politics, Rorty resembles Popper, except that Rorty turns historicism into a good thing, philosophically speaking: "This historicist turn has helped free us, gradually but steadily, from theology and metaphysics – from the temptation to look for an escape from time and chance."[22]

Although Rorty did not develop his critical analysis of scientific knowledge in historicist terms, it is significant that Rorty's major work, which shaped his reputation, is his book on Descartes entitled *The Mirror of Nature.* It is an extended, well- developed attack on the idea that modern science can tell the truth about physical reality. Like Popper, Rorty begins with the denial that the methods of modern science can determine the truth about the universe, its putative object of study. From here, the next step is to deny the validity of any social or political doctrine. "The ironist . . . is a nominalist and a historicist. She thinks nothing has an intrinsic nature, a real essence. So she thinks that the occurrence of a term like "just" or "scientific" or "rational" in the final vocabulary of the day is no reason to think that Socratic inquiry into the essence of justice or science or rationality will take one much beyond the language games of one's time."[23]

This degree of relativism undercuts the firmly held and explicitly pronounced left-wing beliefs of Rorty, which would seem to leave him not merely in a quandary but facing a wall of inconsistency. There seems to be no escape or intellectual work that can serve to justify, in the face of his own philosophical relativism, Rorty's firm commitment to social progress in the contemporary Liberal sense. How does one justify civil-rights legislation if there can be no valid ideal of equality or racial justice to underlie it intellectually and to motivate it morally in the public square? Indeed, how can American society or any society or government survive without an agreed-upon general sense of what constitutes good policy and good social behavior? Rorty insists upon the validity and, by implication, the enforcement of a progressive agenda despite such questions, and terms his philosophy "liberal ironism." As could be expected, it seems that Rorty could never advance beyond this point intellectually, which at a later point in his career caused him to abandon political philosophy (and philosophy altogether) to write and think about literary forms and criticism. Nonetheless, his account of ironism has reached a general audience of informed people and has been much discussed, even to the point of being the subject of jokes, for instance, about ironing laundry and ironists.

The ironist concept has now become a meme in the culture, due not only to Rorty's particular influence but also because the meme reflects a general pattern of decay of optimism among the progressive elites. This occurred, first, about the effectiveness of government programs such as the Johnson era antipoverty programs, and, second, because, as pointed out in the presentation of Whig progress, of the presumed alliance between the progress of science and social progress. The latter issue is very important, since from the beginning of the Enlightenment historical era, there has been an implicit understanding that there is a link between the scientific progress and social progress; discoveries and theories of modern science were presumed to support the social progressive idea. The implied materialism of modern science was supposed to eliminate religious belief and received traditions, which were formerly seen as impediments to the development of a progressive society. But recent experience with the failures of government programs, as well as discoveries in cosmology, biology and in social science, have served to counter the progressive idea that social mores and human nature can be transformed and so made to fit the progressive ideal.

Notes

1. For a detailed account, see John Caiazza, "The Counter-Revolution in the Philosophy of Science," in *The Disunity of American Culture*, 87–102.
2. Ernest Nagel, *The Structure of Science: Problems in the Logic of Scientific Explanation* (New York: Harcourt, Brace and World, 1961).
3. Norwood Russell Hanson, *Patterns of Discovery*, (London: Cambridge U, Press, 1965) 19.
4. Giambattista Vico, *The New Science of Giambattista Vic,* trans. Thomas Bergin and Max Fisch (Ithica, NY: Cornell University Press, 1970) 283.
5. Kuhn, *The Structure of Scientific Revolutions*, 10.
6. Margaret Masterman, "The Nature of a Paradigm" in Imre Lakatos and Alan Musgrave editors, *Criticism and the Growth of Knowledge* (New York: Cambridge University Pres, 1970) 59–90.
7. Isaiah Berlin, *Vico and Herder: Two Studies in the History of Ideas* (New York, Random House, 1976), xvii.
8. George Wilhelm Friedrich Hegel, *Hegel's Philosophy of Mind,* trans. A.V. Miller and J. N. Findlay (Oxford: Clarendon Press, 1976), 293–316.
9. Berlin, xvii.
10. Israel Scheffler, *Subjectivity and Science* (Indianapolis: Bobbs-Merill, 1967).
11. Ibid., 12.
12. Ibid., all quotes 15–18.
13. Ibid., 19.
14. Karl Popper, *The Logic of Scientific Discovery* (1959; reprint, New York: Routledge, 2002).
15. Karl Popper, *The Poverty of Historicism* (New York: Harper and Row, 1961).
16. Karl Popper, *The Open Society and its Enemies* (New York: Harper and Row, 1962).
17. Peter Singer in 1974 on Popper: http://www.utilitarian.net/singer/by/19740502.htm.
18. Popper, *Poverty*; p. iv.
19. Ibid., 3.
20. Friedrich A. Hayek, *The Constitution of Liberty* (Chicago, University of Chicago Press), 1960).
21. On Popper and Thatcher: http://www.nytimes.com/1994/09/18/obituaries/sir-karl-popper-is-dead-at-92-philosopher-of-open-society.html.
22. *Contingency, Irony, and Solidarity* (New York: Cambridge University Press, 1989); xiii.
23. Ibid, 74-75.

6

Where We Are Now: Technology and Culture

The effects and omnipresence of technology in our culture are well-known. What is being referred to here as "*technologization*" is under constant scrutiny, on blogs and websites on the Internet (quite appropriately) and in a steady stream of books, so it is not the intent here to add one more analysis of the same general kind.[1] Nevertheless, it is worth reconsidering the dramatic effects and omnipresence of modern technological applications these days because it reflects the pause in the Whig vision of scientific and social progress. That is, in considering the thesis we are pursuing of alterations in the vision of Enlightenment progress, in social fact as well as philosophical theory, progress has come to a partial halt. Further, to advance the thesis, this perceivable halt is either the result of or is reflected by recent developments in technology in Western culture. Contrary to the usual perception, technology has not speeded up the culture or life in general to the point that there is no longer a floor beneath our feet, so that all that was certain melts into air. Rather the contrary may be true, that technology has had and is having the effect of slowing down change in a cultural sense, and in this way either causes or reflects the recent perceivable halt in progress. There are other complications as well.

We still believe in progress of a sort but it is an attenuated vision, principally because while science does seem to proceed in new discoveries and applications, we have not reached that projected end of social completion and fulfillment, the "broad, sunlit uplands" that Churchill foresaw. We take for granted that progress now means not a general process of cultural enlightenment based on scientific discovery, but rather is exhibited as merely an increase in the technological means of living a comfortable, middle-class life. We are impressed not so much by relativity theory than by the amassment of technological gadgets: the GPS, microwave ovens, handheld computer-phones, video games, computer apps, digitalized movies, thinking thermostats, etc.[2] The total effect of recent technological applications and their impact, that is, technologization, can be seen in several areas.[3]

Techno-Secularism

Among the recent developments in Western culture, secularism is undoubtedly on the rise, and religious faith and the influence of Christian denominations in public discourse has declined. Technologization has been an important part of this: the phenomenon of "techno-secularism" that I have explored means that the effect of technology has been to suppress both religious belief and an appreciation of science itself.[4] Technology has become, in effect, a form of magic. Its effects have been so completely developed and its inner mechanisms so enhanced that it operates seamlessly, providing not only relief from the burdens of life, but enabling us to live in a form of luxury. As a result, we have become intellectually lazy, no longer wondering about laws of nature or the role of experiment as aspects of science, but accepting science's technological improvements as our due. The effect of technologization has had even greater deleterious effects on religious belief, for we no longer look to the Lord or a transcendental vision of life (and the afterlife) for salvation, or to reach a full realization of our human potential. Rather, we look to technology to provide the medical miracles and the creature comforts that make our lives more than tolerable but pleasant. In so doing, we slide past consideration of the big, existential questions that bedevil most serious thinkers, failing to live according to the Socratic maxim an examined life that is worth living.

Yet, the vision that as science and enlightened attitudes progress, reliance on the mythmaking properties of religious faith will decline has failed to be realized. Instead, it has been emphatically challenged by the rise of Islamic fundamentalism and its unrelenting program of terror. Adherents of this religious affiliation attack whatever targets in the West are available, dialing up their ruthlessness and demonstrating their inhuman cruelty in order to make their statement as a kind of reverse progress that leads to the atavistic tribalism of pre-civilization. This assault from the Dark Ages has required a response in terms of self-defense, if nothing else, and most prominent in that defense so far has been the technologization of warfare. In order to deal with the challenge of Islamic imperialism, impotent since the end of the First World War but now an unavoidable presence on the world scene and a specific challenge to Western values and the lives of our citizens, we have relied instinctively, as it were, on our superior technology. So immense is the impact of technology on warfare that it is possible at times to avoid the sense that the West is in fact fighting a war, that is, fighting deadly violence with deadly violence.

The basic issue is the projection of force across continental lines, is how to attack the enemy when the enemy is not next door, as when France and Germany fought each other. Islamic imperialism relies on training its warriors to infiltrate the West and to commit acts of terror. The reliance of the West in general, and of the United States in particular, on drones and aircraft carriers with warplanes represents the very apex of Western and American technology—to attack the enemy from afar. This technologization, in a manner, insulates Western peoples from the fact of the very active warfare going on between the West and Islamic imperialism. As in the case of drone warfare, the warrior's tool is not a sword or a rifle but a joystick that is used to guide the drone from 5,000 miles away, its deadly explosive effects appearing as only as a brief flash on a video screen. When they bring the war to the enemy, the fighting men and women on the ground in Muslim countries have a full complement of technological means to protect and assist them, including Kevlar vests, bomb-sensing devices, efficient assault rifles, and improved medical techniques, (according to a professor of nursing, the last is the only good thing to come from war).

But technology cannot be a means of fighting the "war against terrorism" on the ideological front, or in the realm of the battle of ideas. We are forced to examine the links among the theology expressed in the Koran, Islamic commentaries, and recent developments in Islamic thinking. To what degree, if any, is violence and warfare permitted or encouraged, in the Koran and the Hadith? Is there is an essential link between religious belief in general and the use of warfare and reliance on violence? We need only look to the 4,000 years of recorded history of religious conflict to note its effects on human behavior, the constant irruptions of interfaith warfare, and the heightened cruelty that seems to accompany it. Is monotheistic religious belief the primary—if not the sole—source of intensity that is the cause of religious warfare, with the Eastern religions of Buddhism and Hinduism being more pacific? And finally, what are the sources of religious belief? How can it be that in the twenty-first century it persists like a birthmark of human nature that seems ineradicable and engrained more than on the skin, down to humanity's very bones? The irony here is that as Western culture affects to dispose of religious belief in the public square, we must at the same time engage in what is fundamentally theological inquiry. We are not so very far from the days 400 years ago when a fundamental issue of warfare between Protestant and Catholic was the status of the Eucharistic bread and wine, whether just symbolic or real by means of transubstantiation. Western culture cannot depend on technology to answer questions of what amount to comparative religion and monotheistic theology.

Technology and Cultural Stasis

Popular culture gives us the "same old same old," with new pop stars replacing the old set (Mylie Cyrus replaces Lady Gaga who replaces Madonna). Perversely, in pop culture electronic technology has a conservative effect. Elvis Presley and Frank Sinatra are alive as ever, electronically speaking. No one will do an updated version of Bing Crosby's "White Christmas" or Roy Orbison's "Pretty Woman" as long as those recordings exist, as they now do because the old seventy-eight-speed records have been digitized to reside permanently on disc and in the cloud. The same effect permeates classical music; consider symphony-orchestra conductors as they look over the score of Beethoven's Fifth Symphony in anticipation of a performance. They must now contend with the fact that a goodly number of recordings of the famous work are readily available. How will conductors put their own stamp on the great music, when recordings of "the Fifth" that were conducted by Sholte, Bernstein, and, going back to the 1940s, Toscanini, are available?

And what of the actor who, as he reaches the pinnacle of his profession, has been awarded the role of Hamlet; how will he play it and interpret the character, perhaps for a new age? Unfortunately for him, there are a goodly number of Hamlets available since the famous actors who have previously performed the role are not dead, at least not electronically. In the past, once a stage performance was given it passed into the air and people's memories, but no longer. If the play or the performance is exceptional it has been recorded; the Hamlets performed by Gielgud, Olivier, and Cumberbatch are all there for comparison. Obviously, the best advice for the new Hamlet is to ignore all the others, and to not even watch their performances. To do so would likely make him too self-conscious to perform well.

And then, looking at the conservative effects of technology on culture, there are the movies, or, rather, the movies and the Internet. Movies as originally produced, are in fact only semipermanent because the cellulose on which they are recorded grows brittle over time. Therefore, many early films, such as the comedies of Buster Keaton and the Keystone

Kops, have been lost. However, now movies—or the most famous or important of them—have been digitally recorded and, in some cases, "remastered" to restore their original look. Electronic technology has assured their permanence, at least for the next century or so. (The rates of decay of electronic materials on disks, servers, etc. have not yet been determined, but such decay will take place.) Technological rescue for the movies has had another effect besides permanence; because they are digitalized, films are now readily available on disc and on the Internet. Movie buffs no longer have to go to art houses that specialize in offbeat and foreign movies to see *Battleship Potemkin, The Searchers, Singing in the Rain, Bringing Up Baby*, or *The Third Man*. This, in turn, has led to the development of a virtual community of movie buffs whose members are their own film critics and historians; every movie buff his or her own Pauline Kael or Roger Ebert, communing with one another on websites and blogs.[5] This phenomenon has a distinctly conservative effect, for the past has become ever present, likely crowding out new developments an ideas. Why waste your time with a postmodern take on the spy-action genre by an unknown director when you can rerun Hitchcock's *North by Northwest*?

Watching old movies reinforces the point that one of the contradictory effects of recent technology is to make the culture static. When watching a film from the 1940s such as a crime film or a romance, we may at first be aware only of the contrasts and improvements. The rotary telephones are not slim, portable and handheld but large, clumsy instruments that must stay in one place, the cars are boxy and not sleekly designed, etc. Yet consider what is the same between what the movie betrays was the case then and what is the case now. Telephone communication is basically the primary means of talking with someone you can't see because they are too far away, and that has not changed. The phones in the 1940s may have been clunky, but the grim detective or the romantic heroine could still make a call from Los Angeles to San Francisco, or from Los Angeles to anywhere else on the globe, to speak with someone in real time. The cars may be old-fashioned but they were nonetheless powered by internal combustion engines and were driven on cement or asphalt streets. The detectives dealt with witnesses and potential villains in business offices in which secretaries and receptionists were in front with their bosses in the backrooms, and elevator buildings were skyscrapers built of steel, stone, and glass. The houses and hotels in which the heroine met her lover or rejected her tiresome suitor look very much like the houses and hotels we live in now; in some cases they are the same.

Science Itself

But we are concerned with more serious matters than culture and the movies, or so it seems. As for science itself, it may seem as if its progress can be understood apart from the social context of ennui and halt. We can now look at the remarkable color photographs from deep space that are provided by the Hubble Space Telescope, which can see far enough into the astronomical distance to reveal the very first generation of galaxies. We can read up on the latest experimental results of the Large Hadron Collider (LHC) that is searching for the "God particle." Yet, it has to observed how "big science" because of its dependence on big funding, must have some appeal to the general voting public. After all, the billions spent on the Supercooled Supercollider could reasonably have been spent elsewhere, as the US Congress determined when it shut down the American version, which would have buried several miles of vacuum-sealed pipes surrounded by a series of powerful electromagnets under the ground in Texas. Like a Defense Department boondoggle, after its estimated cost had immediately doubled, it still cost over a billion dollars to shut down.[6] The project however was completed in Europe and was funded by a consortium of nations (CERN).

But what appeal could be made to governments and peoples in a democracy that would be convincing? Why worry about the opinions of physicists who seem to need vast amounts of public money just so they can uncover yet another subatomic particle of which there seems to be a plethora already? The primary popular argument for the LHC was that it would allow the discovery of the Higgs Boson—the God particle—and it well may have been the resonance of ultimacy, of physical finality linked with divinity, that carried the day. Who would be churlish enough to deny the physicists their funding when the GOD PARTICLE was just next door, waiting to be observed as the door opened on the experimental proof of the "standard model" of subatomic reality, the final resting place of the Reductive ideal of progress. What could make more sense than an ultimate scientific explanation that rests on an ultimate particle? However, the (possible) discovery of the Hoggs Boson is the (probable) confirmation of an already widely accepted theory that has been in existence for over two generations, and has therefore not provided much new content to physical theory.

The Hubble Space Telescope has given us more than just colorful images; it has opened a window for the scientific analysis of the most primitive and ancient origins of the stars, elements, and galaxies. But it is the pictures that compel popular interest. While the revelations provided by the Hubble telescope do tell us new and exciting things about the universe, the discoveries seem problematic in that their primary revelations about deep space is to imply the existence of dark matter and dark energy, which only increases the mystery of the universe.

It may seem inappropriate for philosophers to predict what the future of scientific research holds, and certainly it does for the scientists themselves. Yet, prominent scientists have had no reluctance to discount philosophy as a fruitful intellectual activity, and to predict the decline of religious faith. Turnabout is fair play, especially as we consider how recent technology has in a way turned the manifest progress in basic research back into philosophy. This is a point that will be developed further in the subsequent chapters. But here we can point out that as the technology that underlies scientific experimentation has progressed, it has revealed certain sets of facts that do not so much contradict dominantly held theories in particular fields as they do, so to speak, *complexify* them. That is, in physics, the ongoing discovery of a number of subatomic particles has meant that theory has had to keep up. An example of this can be found in chapter 2, which discusses how spectrographic analysis of the stars revealed not only their internal composition but also their paths away from each at increasing speeds. This was a discovery of fundamental importance for understanding an essential property of the astronomical universe, namely, that it is expanding.

But expansion, and now inflation, does not, as we stated, so much contradict as make more complex the state of scientific/physical theory. It is not only that a theory must be improved by the addition of a subordinate clause or a dependent piece of deductive reasoning to include some new discoveries. Instead, at the fine-grained level—at the material bottom of what a particular field studies—theories must become so complex and so rarified that they may lose logical coherence. In physics, the advances in astronomical discovery and atomic experiments are not only reflected in the development of new theories, but relativity and quantum theory have become in a logical sense irreconcilable. Such a development turns physicists into philosophers who are forced to consider the role of the experimental observer when looking at the largest and the very smallest of natural phenomena. Physicists now realize that there are serious, if formerly overlooked, questions about the relationship between the human intellect and the material universe. In this manner, at the leading edge of physics and in scientific discovery generally, technologization has led to a complexity that cannot be cleared up simply by means of a new or improved theory.

The same effect of "complexification" can be seen in recent developments in evolutionary biology. Discoveries that have come about through genetic research contradict the idea that is basic to Darwinian evolution: that evolutionary change must come about very slowly. The fossil record could never readily support the idea of species change gradually coming about over tens and hundreds of thousands of years, but now genetic explanations and discoveries make that concept even more problematic. As a result, many additions have been made to the classical theory of evolution by variation and natural selection in order to account for what genetic science, enabled by technologies such as the electron microscopes, X-ray crystallography, and historical comparison of mitochondria DNA, have revealed. Despite recent claims made in popular books, at its fundamental level evolutionary theory is not reaching a state of triumphant explanatory success. Rather, it has become so complexified that it ranges into philosophic concerns, such as the relation between vast lengths of time and the ability of the human intellect to make inferences. Other such concerns are the complexity of biological processes themselves, and how random biochemical activities can explain the development of increasingly complex species over time.

Digital Fantasy Replaces Lived Reality

To summarize the effects of technologization in a social sense is to say that it results in replacing reality, in other words, the lifeworld, with a photoshopped image of the real thing. Indeed, our lifeworld has become what electronic digital applications and the mechanical easing of the burdens and inconveniences of life imply that it is. Common examples are sometimes the best.

Football is a game that boys and young men play. But for the majority of sports fans, with its collisions of bodies impelled by a quarterback's instant command, it is not something they participate in. Fans generally do not attend an actual game out of doors, where the cheers, disappointments, and excitement are experienced in a crowd that is sharing the same reactions; rather, the game is watched indoors on television. Thus, it is possible for a Californian to become a fan of the New England Patriots or a New Englander to be a fan of Notre Dame. Watching football on a flat (and wide) television screen indoors does not of course replicate watching or playing the game, for it is a process of merely watching moving pictures on the screen. So much is understood, but the process of watching football in the age of digitized electronics has now proceeded to a further stage. This stage not only further removes the football fan from the real thing, but replaces reality with desire; the term "fantasy football" carries with it a potent accuracy.

Fantasy football gives the sports fan the opportunity to select players for his or her team from any team at all, so that the best linemen, the best defensive back, the best quarterback can be chosen. The selected team is than matched against other players' teams and a "game" proceeds. But it is not a physical game at all, for it takes place as the result of a computer algorithm in which the odds are placed against each team and then randomly run. What the fantasy football participant sees on the screen is not the broadcast of an actual game, but a simulacrum of a game in which his or her selected players come against those of another player. Digital action figures make blocks and tackles or throw the football, with the outcome dependent only on the statistical results. However, reality is in a way reintroduced into the process since fantasy football is connected to online gambling, so winning or losing between fantasy teams becomes a matter of winning or losing money. In this manner, fantasy football gives the participant a real stake in the game, but it is a fantasy game that distorts and alters, and, we may say, even replaces reality. It is not only that a sweaty, physical, and sometimes dangerous contact sport is replaced by computer-generated statistical odds and electronic images. It is that the total experience of watching or playing a game of football has become something significantly different, a replacement of the reality of watching or playing the sport with a complex computer game.

Fantasy football is an extension of the digitalization of popular home entertainment, where we see another negative effect of technology on culture: the replacement of outdoor play by indoor video games that young people and children can engage in while sitting down. The unhealthy results have become so widespread that children's programming now contains encouragements to get up and walk around rather than sitting down all the time in front of a video screen. The digitized experience available to children is one of the causes that have been cited as resulting in a new phenomenon, childhood obesity. But there is a social distortion as well. The video version of fantasy football, for example, makes the player an entrepreneur, a king, and a kingmaker—the sole determinant of who wins and loses, what the make up of the team is, who the stars are, etc. But real football is a team sport that requires, even more than baseball and soccer, a close coordination of the simultaneous actions of all the players on the field. In fantasy football, team aspect is replaced with a narcissistic illusion of total control, which, in psychological terms, may be more destructive to the individual than obesity. Lack of social interaction makes one eventually unfit for society. (Interestingly, there may be some recent objections to the trend of isolated video-game playing. This is apparent in the recent resurgence of the popularity of board games played among small groups of people, with the ancient Monopoly replaced by such updated versions as Catan and From Trails to Rails.)[7]

The Electronic Ego

According to the reality principle, so described by Freud, the desires of the id are frustrated when the ego is presented with unalterable facts. Such a fact is the awareness that a healthy sense of life is achieved when we interact with others whose desires and actions are as real as our own. This requires accounting for the needs of others, an action that in turn alters our personal behavior. In Freudian terms, the reality principle is related to the development of the ego, the behavioral "face" that we present to the world and by which we satisfactorily function as an individual in society. Reference to the reality principle is a useful means of designating what happens in human experience when lived fact replaces fantasy, when young girls learn there is no special place where pink unicorns live and young boys may realize that they lack the aggressiveness to make the football team. Such replacement of fantasy with reality is usually seen as a process of maturation, of growing up, or, at its worst, of substituting idealism with cynicism. Young girls learn to socialize and to regard peers and others as individuals with whom it is pleasant and worthwhile to get along, and young boys may turn to another sport such as gymnastics or chess in which they may succeed. Now, it should be stated that fantasy is not a bad thing. It can be useful in making plans for the future, and in setting up goals and projects for ourselves; by imagining the future we have something to aim for and can give structure to dreams and future activities. But looking at the example of fantasy football and its effects, we now pursue the question of *what happens when the reality principle is denied in this fashion so that fantasy suppresses reality*?

The answer is that technologization provides a kind of virtual isolation chamber, in which the lack of interaction with other people and of an external or "lived" reality makes the "I"—the ego—the sole and unassailable center of the universe. This, in turn, makes the development of a sense of social ethics impossible, a fact that is reflected in a lack of implied ethical content in the technologized society. Contemporary ethical thought is usually based on two general types of theory, utilitarian and deontological ethics, neither of which is supported by technologized culture. The ethics of technologization is not utilitarian, as may be expected, for it is not the greatest amount of happiness for the greatest number that is implied. While it may be said that utilitarianism lacks a transcendent horizon, it is nonetheless a social ethic that demands a response to the unjust treatment of certain classes of society including the poor. In contrast, technological ethics aims at the individual by providing sustenance for the individual ego, but also by suppressing the ego's sensitivity to external reality and to the reality of other people—perhaps the reality that is the most difficult to recognize. But, as with utilitarianism, technological ethics recognizes only the material realm and does not extend into the transcendent (much less the religious). It is a practical materialism that has the further aspect of denying not only the reality of the transcendent, but the reality, in effect, of society and of other people. Here then, we can see that technological effects provide no basis or grasp on the reality of other people, and that, in deontological terms, in terms that is of a universal duty to others, self-regard has made the creation of a coherent social ethics impossible from the start.

In terms of contemporary ethical theory, what technologization promotes is egoism, an egoism that is somewhat like that of Nietzsche for it promotes an inflated sense of self-importance on the cosmic level. The ideal of the "superman" is seen in the video games in which one becomes an all-conquering hero and is given such names as "dungeon master" and "great wizard." It is adolescent, almost childish, and leads to difficulties of social adjustment if such identification is not outgrown. If we presume that the ideals of social ethics arise from the experience of social interaction, then it can be seen that the technological "I" is foreclosed from the basic experience of social interaction from which as a practical matter social ideals and concepts of appropriate behavior arise. But beyond the confines of a boy's bedroom and video gaming, what are the effects of technologization on the adult world?

Here it is noticeable that technology is put at the service of middle-class comfort, and that the products of ever-advancing technology are sold to consumers on the basis of convenience and pleasure. In the 1920s the Hoover vacuum cleaner, mechanical washing machines, and so-called "ice boxes" were new, labor-saving items that did not so much make living more comfortable, but made living possible on a middle-class level. In terms of time and energy, it was liberating for housewives not to have to clean rugs by hanging them on a clothesline and bat at them, not to have to take dirty clothes to a river bank or common water trough to literally beat them clean, to be able to refrigerate food so as to keep it for more than a day without spoiling. Such technological assists are replicated today but with much less improvement or effect on practical living. How does a new-model vacuum cleaner that uses a cup instead of a cloth bag to collect dirt increase utility or save time and labor? So, also, what improvement is gained by a water-saver washing machine or a side-by-side refrigerator? Such improvements are so minimal and their advantage so limited, that more money is spent by manufacturers in advertising these new items than in manufacturing them. What difference does it make if one owns the third iteration of an "iPhone" instead of the fifth? The total effect is that the average consumer is made to feel as if they are personally and solely the object of technological advance. Despite the fact that these products are sold to millions, the advertising, logistics, and sales processes are devoted to the idea that the single consumer has an entitlement to these new products. Once again, the technological "I" has become nowadays the center of technologized commerce.

The most destructive and tragic consequence of the technological replacement of life experience with digital imagery is pornography. While this is a disturbing topic, it is not avoidable because digital technology has made pornographic images omnipresent and immediately available on the Internet. All the negative effects of current technology described so far in this chapter are intensified. In the past pornography was only available as hard copy which made it much harder to find and easier for authorities to control. The degradation of both women, whose images are deformed and made into sex objects, and men who often become addicted, is only too well known. Overall, the effect is that desire is frustrated and social interaction between the sexes becomes more difficult than it normally is. False understanding and doomed expectation between men and women begins with the shadow images available on-screen and on demand 24/7. The much expanded availability of pornography is the worst effect of the new digital technology.

Notes

1. Ulrich Beck, *Risk Society: Towards a New Modernity*, trans. M. Ritter (London: Sage Society, 1992).
2. Nicholas G. Carr, "IT Doesn't Matter", *Harvard Business Review*, May 2003, https://hbr.org/2003/05/it-doesnt-matter.
3. Technologization also has a profound effect on practical ethics, the primary effect of which is the separation of actions from consequences, for example, the development of life-saving technologies such as air bags makes poor driving less hazardous to one's health. I have written about technology and ethics in *The Disunity of American Culture* and in *The War of the Jesus and Darwin Fishes*.
4. John Caiazza, "Athens, Jerusalem and the Arrival of Techno-secularism," *Zygon*, 40, No. 1, March 2005.
5. See for examples of virtual communities of movie buffs http://facebook.com/indy-filmbuff, and www.imdb.com/list/Is057552858.
6. See Weinberg, *Dreams of a Final Theory,* where the episode of the supercollider built in Texas, which was never completed, is recounted.
7. Information given to the author by Martin Ammer.

7

Philosophy, Progress, and Cosmology

Progress having been critically examined from the point of view of providing a description of the advance of scientific theory and in social terms, we can reexamine the case for progress in more intellectually inclusive terms, that is, in philosophic terms. By now it must be admitted that modern science has, at its forward edges of experimentation and theory, entered into philosophic realms. This chapter will provide an interpretation of what scientific progress looks like in the philosophic time line, wherein science first departs and now returns to its philosophic origins. In this interpretation, modern science (having departed from its philosophic origins in the sixteenth century—or so it has been claimed and widely accepted) now returns, in the manner of completing a historical arc, to its philosophic "home." The primary example of this phenomenon is the arrival of contemporary cosmology as both a scientific field and ground for philosophic analysis and speculation.

Modern Science and Philosophy in Contrast

From the beginnings of modern science, its advocates, such as Bacon, the positivists of the nineteenth century, the logical positivists of the twentieth century, and such current figures as Krauss (see below)—have made the argument, or, more accurately, the assertion that science supercedes philosophy. They have contended that philosophy is woolgathering, and armchair speculation that reaches grandiose conclusions absent evidence for them. This characterization is contrasted with the exact and empirically founded laws of modern science, which is viewed as having far more practical use for the human race than the multivolumed mental effusions of Plato or Kant. At best, it may be argued on behalf of philosophy that it discovers problems and formulates basic questions that have intrigued mankind from the beginning, but that science answers those questions thus closing off the need for further speculation or consideration. In the terminology of one philosopher (Marx Wartofsky), philosophy serves as a propaedeutic to science and merely serves up the intellectual challenges for which modern science can provide the answers.

This notion results in controversy, as recently illustrated by Rebecca Goldstein's *Plato at the Googleplex: Why Philosophy Won't Go Away*, which begins with a direct response to another recent book by physicist and pop-science writer Lawrence Krauss.[1] Krauss asserts that because science now has the answer to the ultimate big question of the origin of the universe, whatever is said by anyone else from a philosophic or religious point of view is insignificant. Krauss relies for this hubristic claim on recent developments regarding quantum energy fields, from which it is alleged physical particles pop up into existence as if from a vacuum. Krauss's argument involves the definition of "nothing," and whether quantum fields of energy are quite what Augustine meant in his gloss on the first lines of *Genesis*, namely, that time itself began with the act of Creation, before which there was *nothing*. The subsequent debates with philosophers and religious critics remind older television viewers less of Einstein and the big bang theory than of "Seinfeld," a show "about nothing" since nothing seems to have been settled. Physicists nonetheless seem happy with this formulation, easing as it does their imaginations from the prospect of the big bang, before which their laws and mathematical formulations do not apply. Further, the formulation is uncomfortably reflective of the biblical account of creation *ex nihilo*, that is, "from nothing." Krauss himself has become a personality, having joined the parade of scientific promoters of atheism that include Hitchens and Dawkins.[2]

The author of *Plato at the Googleplex* has her own ideas, for she thinks that the Socratic pronouncement "the unexamined life is not worth living" applies to theoretical physicists and evolutionary psychologists as much as to philosophers and ordinary people. Goldstein makes her case by portraying Socrates in dialogue with people in different contemporary settings, including a right-wing television talk-show host and the computer engineers at the "googleplex" (the corporate headquarters of Google). Goldstein's arguments, which replicate those of Plato in his dialogues, are extended, detailed, and well-done. They demonstrate, according to her intent, the continuing relevance of philosophy as expressed by the most influential philosopher in Western culture. At this point in the debate about whether modern science can replace philosophy, it is appropriate to ask for a description of philosophy in comparison with science so that we can see what otherwise would be missed if philosophy were to be discounted. From its start, as it first appears in Western civilization and elsewhere as in ancient India, philosophy has had these five characteristics: it is *rational, abstract, ultimate, dialectical,* and *arises in crisis*.

Philosophy is *rational* in that it proceeds and reaches conclusions by means of human reason alone—independently, that is, of tradition, myth, religion, or public opinion. Science applies a skeptical methodology that tends to eliminate the same external influences as philosophy. But philosophy does not have the predilection to physical and mechanical explanation that science does, and so it encompasses a much wider range of types of explanation, for example, teleological.

Philosophy is *abstract*; in fact, it is the most abstract and general of all possible intellectual approaches, a characteristic that modern science shares with philosophy. However, the general propositions of philosophy are not usually based on direct empirical evidence as is the case with the low-level laws of science. Nonetheless, there is an evidentiary basis for much philosophy even at its most speculative, for it does not limit its sources of information to the results of physical experiment. It includes intuitions of the nature of existence, artistic impulses, and reflections on the desires and beliefs of the common people, whom it does not despise but includes as part of its underlying structure of justification.

Philosophy *seeks ultimacy* in that its examinations and explorations attempt to find final explanations. It does not pursue final causes necessarily, but attempts to settle all questions in regard to whatever issue is at hand; a comprehensive skepticism settles issues as well as does a doctrine of final causes (Hume vs. Aristotle). Contemporary mathematical physics also is making attempts at ultimacy in its well-publicized search for the final theory of everything, the Grand Unified Theories and Theories of Everything that reflect Newton's first attempt at an ultimate scientific theory in his *Principia*.

Philosophy's main method is that of *dialectic*, that is, a conversation or debate intended to identify, set forth, and subject to criticism certain positions. The goal is to reach a definite but not necessarily final conclusion. Modern science has a corresponding methodology based on continual empirical testing of its laws and discoveries, and so attempts to be self-critical in that regard. But science also tends to be more conservative and less self-critical than philosophy in that its major theories are usually accepted as given and as the basis for further research.

A last characteristic of philosophy as I have outlined it for my students in introductory courses is that it *arises in times of crisis*. That is, philosophic inquiry is often forced by a practical necessity of thinking human beings to understand their circumstances when their circumstances are radically changed. The fact is that anyone of normal intelligence, but especially if well-educated, can do philosophy, for it arises spontaneously whenever people come to wonder about things on an abstract level. It's not just Plato and Rawls who do political philosophy; politicians like Burke, Jefferson, and Lincoln who assess the political difficulties of their times by reaching for a general idea from which to justify and clarify their policies are also doing philosophy. This is also true of scientists who, when they encounter difficulties, often reach for the general principle or a means of rational understanding that goes directly into philosophical areas.

Three Examples of Scientists Doing Philosophy (and Theology)

Here are three examples that indicate that however distinct science and philosophy are claimed to be, major scientists often have recourse to philosophic thinking at times of crisis. There are many examples but these are among the foremost.

Galileo on Biblical Interpretation

Galileo's conflict with the Catholic Church is the foundational myth of modern science; it is, however, a true myth from which several lessons may be drawn. One is that modern scientists are often driven to do philosophy and, in Galileo's case, theology in order to explain the ramifications of their discoveries and theories. Galileo's advocacy of the Copernican system resulted in the accusation of heresy, from which Galileo as a personal matter absolutely needed to clear himself. Being charged with heresy in a church court could result in imprisonment, as it did eventually for Galileo, and fiery execution, as it had two decades earlier for Giordano Bruno. The charge had not been officially made (that was to come later), but was first made by a Franciscan friar who did so publicly in the court of Grand Duchess Christina. Christina was an important personage, a member of the powerful Medici family whose court entertained the influential figures of the day. Therefore, the charge against Galileo became well-known. It was not Galileo's first brush with the charge of heresy; he had several years earlier had to clear himself of the charge by means of a formal conversation with Cardinal Bellarmine, known as the primary defender of Catholic doctrine in a time of Protestant revolt.

The basic issue was that the Copernican doctrine could be interpreted as contradicting several biblical texts that state or plainly imply that the sun circles the Earth.[3] This was the basis of the charge leveled by the friar. But in order to clear himself of the charge of heresy, it was necessary for Galileo to develop his own method of biblical interpretation.

Galileo replied in an open letter to the princess. This was not a personal letter since it ran about 8,000 words and contained intensive arguments and analyses of the situation, mostly revolving around how biblical texts are to be interpreted when they give descriptions of the natural order. The general notion that the Bible ought not to be taken literally in those cases where the evidence of the senses gives a contrary description of nature was not new at that time. Augustine had made the distinction a thousand years earlier, and Galileo quoted by name the Cardinal Head of the Vatican Library who said, in a little ditty, "The Bible doesn't teach you how the heavens go. The Bible teaches you how to go to heaven."

The general point Galileo made was that the Bible uses language that fits the understanding of the people of ancient Israel, a Semitic tribe in the ancient Middle East whose understanding of the natural order was limited. The Israelites were at that time a primitive people whose culture suffered in contrast to that of the Egyptians, the Babylonians, and the Greeks with whom they interacted. What Galileo emphasized in his argument, however, was that the Bible used figurative language even in its description of God. Christian doctrine, supplemented by Greek philosophy, described God as an immaterial and purely spiritual being. But Galileo pointed out that certain texts of the Bible gave God human, physical attributes such that he became angry and pleased, and referred specifically to God's hands and feet.[4] Obviously, the Bible was using figurative, not literal, expressions in order to make essential points about how human beings ought to regard the Supreme Judge and supreme metaphysical being of the universe. The conclusion, in Galileo's mind, was obvious: If the Bible utilizes figurative language to describe the nature of God, the Bible's primary focus after all, why must its descriptions of natural order be taken literally?

This is a question of current import as well. Currently, Christian fundamentalists state that the Bible must be taken literally, and on this basis reject the theory of evolution. They have taken their opposition into the public schools. As a result, there have been several well-known court cases that have pitted defenders of science against Christian believers in the literal interpretation of scripture over the issue of what should be taught in biology classes in local public high schools. While defenders of evolution have so far won these battles in the courts, nonetheless the opposition has been somewhat successful in promoting an alternative to evolution, namely "intelligent design" theory. That theory is claimed to be false, but defenders of evolutionary theory have also been at fault in this controversy, that is, in promoting evolution as a materialistic replacement for religious faith. The assertion that evolutionary theory implies a materialistic metaphysics has been a constant element of presentations of the theory, starting with Huxley and Spencer and extends currently to Dawkins and Dennett, who have made evolution the basis for an outright attack on the Christian religion.

This aggressive attitude is in contrast to Galileo's during the earlier controversy. Currently, evolutionary materialists have exhibited a search-and-destroy mentality toward traditional religious faith. This is in direct contrast to Galileo, who was sympathetic to it (being a believer himself), and who attempted to make his case in terms that would be acceptable to religious believers. Is it beyond the intellectual capacity of evolutionists to point out that the account of life's creation in *Genesis* and that of Darwinian evolution share a *developmental* approach to origins, and use this correspondence to make an appeal to biblical-minded Christians? While Galileo's efforts failed during his lifetime, in broad terms his view—not only of the Copernican theory but of the relationship between science and religion—has become the standard view of mainstream Christianity.

Bohr vs. Einstein on Subatomic Indeterminism

The debate between the two greatest theoretical physicists of the last century, a period that rivals the seventeenth century as a time when revolutionary discoveries were made in physics, is often remarked upon. There is a well-known photograph that shows Einstein leaning back in a chair holding a cigarette, staring straight ahead, apparently listening to the comments of his younger colleague, Bohr, who is speaking while sitting straight-backed on the edge of his chair. There is another photograph of the two scientists walking side-by-side, Einstein looking straight ahead again while beside him Bohr, who is somewhat shorter, is speaking energetically. Einstein's straight-ahead stare in these photographs does not mean that he was bored with what Bohr had to say; on the contrary, the two men argued over several decades over substantial issues of the explanatory, conceptual, and literally metaphysical consequences of quantum theory. That is, they debated the nature of atomic substructure, that is the many particles and corresponding energies that lie beneath the surface of the atoms themselves, how they are observed and discovered, what the interrelationships are among them, and how well, if at all, their exotic mathematical descriptions—collapsing waves of probability, matrices—accord with the equations of relativistic physics.

Einstein and Bohr took opposing sides in this significant debate, but their primary means of making their points and contradicting the arguments of the other was by means of "*thought experiments.*" Rather than referring to experimental evidence and mathematical equations, Einstein presented a series of "what if" experimental scenarios that could not actually take place in a laboratory setting, but that he nonetheless claimed defeated one or another of the propositions of Bohr's so-called "Copenhagen interpretation" of quantum physics. Bohr responded with either a direct answer to Einstein's claim or a thought experiment of his own that attempted to rebut or clarify Einstein's conclusion from his hypothetical scenario. Such thought experiments were not new, even as Einstein and Bohr used them against each other's arguments as if they were "light sabers." Shrödinger had presented the problem of the hypothetical cat who, when it left a box observed in the manner of quantum physics, was in one state inside the box and a radically different state when it left the box, with the necessary consequence that being in both states at once would mean the destruction of the theoretical cat. But the climactic example is the famous EPR (named for Einstein and his collaborators, Podolsky and Rosen) thought experiment, which, contrary to Einstein's expectation, resulted in real-world, experimental consequences.

The EPR thought experiment was intended to demonstrate the futility and outright craziness of the Copenhagen theory; if it were true, then two conjoined or "entangled" particles could affect one another even if widely separated and traveling away from each other. That is, if electron ***a*** and electron ***b*** were *complementary* particles, to use the exact expression, then once entangled whatever happened to one would immediately have an effect on the other. If the charge or spin of electron ***a*** were changed, then that of electron ***b*** would change correspondingly and *instantaneously and no matter how far apart they were*. Einstein could not assert that there was experimental evidence that positively disproved the Copenhagen interpretation. But he could show its fantastical and counterintuitive consequences by means of this thought experiment, which he thought he had done. Yet, nature defeats even Einstein, for when it became possible to actually carry out the EPR experiment in a laboratory, what Einstein thought impossible did take place. Once two particles were entangled, a change to particle ***a*** resulted in an immediate and corresponding change in particle ***b***, even when they were far apart.[5]

There are too many issues here to disentangle. However, the main one is the incomprehensibility of atomic structure itself, for it seems impossible to imagine pictorially how the insides of an atom exist or to construct a mechanical explanation of subatomic processes. As a result, the controversy between Einstein and Bohr has never been resolved by physics because it takes place on a plane of abstraction independently of direct empirical data, and is therefore, in a precise sense, philosophical. What discussion and debate about quantum theory involves (as can be seen from contemporary debates about many worlds, alternate universes, and extra dimensions) is, as philosopher Mary Midgely has stated, "metaphysical." These speculations are inherently philosophical, as what the physical theorists are doing is producing metaphysics in the standard philosophical understanding of that term, that is, an overall theory of reality.

Given the hitherto radical separation between quantum physics and relativistic as well as classical physics, a properly metaphysical issue arises: Does the success of quantum physics imply that all of reality is a matter of quantum fields and probability, or does quantum mechanics remain an isolated field separate from the rest of physics? The methodological issue inevitably becomes a philosophic issue: how is physical reality to be uniformly described?

Skinner vs. Rogers on Modern Psychology's True Purpose

By the 1950s it had become apparent that there were multiple ways by which psychologists were doing their science, and that some of them were in conflict with one another. Certainly Freud's version of psychoanalysis, which was influential among the educated public, stood out in contrast to other approaches. But Freud claimed that his version was based on his experience of doing psychoanalysis, which had led him to, among other discoveries, the behavioral effects of suppressed childhood memories. Nonetheless, the experimental approach, carried out in laboratories under standard protocols, its results quantifiable and published in bona fide journals, had a prominent presence in the field. Excluding Freudian psychology, the general breakdown of the conflict was then, and is still, that between practitioners (including psychotherapists, psychologists, and counselors who work with clients to try to help them by relieving their symptoms and advising practical approaches to their problems) on one hand, and experimenters (who work in laboratory conditions to analyze highly defined kinds of behavior with the purpose of explaining them by cause-and-effect relationships—stimulus-response, aversive conditioning, and statistical correlations) on the other. By the 1950s the conflict had become so apparent that it evolved into a formalized debate between each side's major representatives, Carl Rogers and B. F. Skinner, respectively.

In his work with patients, Rogers developed an interactive and personalist approach to psychology. As he noted, "the essence of therapy, as I see it carried on by myself and by others, is a meeting between two *persons* in which the therapist is openly and freely himself and evidences this perhaps most fully when he can freely and acceptantly [sic] enter the world of the other."[6] Rogers called his form of therapy "self-actualization" therapy, in which patients were encouraged and aided to become more fully themselves, apparently meaning that they not only realize their ambitions or succeed in their chosen endeavors, but that in some interior way they become more *real*. "The way to do is to *be*," Rogers stated.

Skinner began his career as a behavioral researcher with an explicitly positivistic understanding of science, looking only to behaviors that were physically present in a laboratory setting and that could be measured and then mathematically analyzed for patterns and statistical correlations. He was famous for working not with human subjects but with animals suitable for laboratory experiments; the pigeons Skinner worked with became famous as he trained them to perform outlandish behaviors such as strutting with their heads and beaks pointed in the air, and for wartime work in guiding aerial bombs. A devotion to objective research led him to discount among human beings their own self-consciousness, eliminating intention and will entirely as causes of human behavior. Skinner looked entirely to training and external conditioning to influence the behavior of the human population, but he preferred positive to negative reinforcement. He extended what can be plainly understood as a philosophy of human nature to society, elaborating in two popular books including a novel how society ought to be organized and controlled. In a formal debate with Rogers that appeared in 1956 in the pages of *Science*, the foremost journal of general science for professional researchers, Skinner made his point:

> It is the experimental study of behavior which carries us beyond awkward or inaccessible "principles," "factors" and so on, to variables which can be directly manipulated. . . . It is also, and for more or less the same reasons, the conception of human behavior emerging from an experimental analysis which most directly challenges traditional views. Psychologists themselves often do not seem to be aware of how far they have moved in this direction.[7]

He went on to explain what this experimental/behaviorist view implied in three areas: the personal, in education, and in government.

Aware after World War II of Fascist and Communist efforts to make over the behavior and belief systems of the populations under their control, Skinner faced a problem: how could he be sure that the social controls he was advocating would not be used for immoral purposes? He asserted that such control would be used only for ethical purposes. In his response to Skinner, Rogers pointed out that all approaches to psychology, whether experimental or therapeutic, begin with a choice of values—a choice that he implied is inescapable. He might also have pointed out the näivety of Skinner's belief that there are certain positive influences in society that guide the people in charge toward proper and beneficial social ends. Such näivety is notable in a scientist who was known for a hard-headed approach to the subject matter he studied.

The actual presence of philosophic expression in the Rogers-Skinner debate would seem to be so obvious that it is not necessary to elaborate on it, especially as Rogers himself makes the point. The presence of philosophic expression in psychology in particular and in the social sciences generally would seem to be a necessary part of doing research into human reality. The search for the origins of human behavior implies a concept of human nature. Skinner avoids such *essentialism* (a term he does not use), but the very vacancy of human nature he assumes, as implied in the title of his popular book *Beyond Freedom and Dignity*, is itself a philosophic position. That is, to say that human beings have no inner reality or free will independent of external influences is precisely a philosophic doctrine.[8]

At the end of the debate, Rogers has the last word—not on whether Roger's humanism or Skinner's experimentalism is the correct method, but on the idea that social scientists ought to try to be more explicit about their "starting values." Just trying to explain the presumptions of their research in terms of the constitution of human nature would have the beneficial effect of exposing the working philosophy of social scientists in all the various fields not only to others, but to themselves. "Know thyself!" should be the first command of social science as well as philosophy.

Is Not Naturalism a Philosophy?

One of the foremost contemporary philosophers of science is Abner Shimony, emeritus professor at Boston University, where for several decades he has imparted to students in the graduate school the outstanding erudition he obtained from his two doctorates: one in physics and the other in philosophy. (Among his students back in the late 1960s was the present author.) At the end of his teaching career, in proper academic style, Professor Shimony published his numerous papers in the book *The Search for a Naturalistic World View*.[9] Yet Shimony, winner of the Lakatos Prize for contributions to the philosophy of science, had early in his career speculated about the metaphysics of Whitehead in respect to subatomic physics, and now at the far end of his career pursues the philosophically vexed issue of quantum *entanglement*.[10] The point is that naturalism is seen as the proper academic bedrock for the remarkably fecund speculations of Professor Shimony, as if reference to Whitehead's metaphysics or any kind of metaphysics in the title would discourage serious philosophers and scientists from reading his outstanding contributions.

The primary philosophic school of influence in America today is *naturalism*, which basically is the idea that nature is the only thing that exists. Nature has been defined by the scientific method, as primarily expressed by a generalized idea of Darwin's theory of evolution. Once accepted, naturalism immediately contradicts the basic thesis of the argument of this chapter that a philosophic history is the best approach to understanding modern science. As modern science departed from medieval philosophy in the seventeenth century, it is apparent that, in the twenty-first century, it must return to its origins in philosophy. Naturalism, once taken for granted, has a deadening effect on philosophic speculation and so directly contradicts the assertion that if modern science is to remain comprehensible to itself and to outside observers, a return to philosophic origins is required.

But naturalism is difficult to define. According to the Stanford Encyclopedia of Philosophy, "The term 'naturalism' has no very precise meaning in contemporary philosophy . . . [and] is not a particularly informative term as applied to contemporary philosophers."[11] Because naturalism is difficult to analyze critically, its expositors have made a distinction between *methodological* and *ontological* naturalism in much the same way as scientific materialism has been distinguished; the resemblance is not accidental. But this distinction only evades and does not basically clarify the philosophy of naturalism. The point here, however, is not to refute naturalism—even as its acceptance invalidates pretty completely the historicist view of science. The point instead is that naturalism does not escape the main idea that is being made in this chapter, that is, to discuss modern science in its present state, given all its cultural and theoretical ramifications, is to be doing philosophy and not science in its classically empirical mode. Thus we can say that naturalism is, in effect, a pretense that implies that within the bounds of modern science, expressed typically in an expanded theory of evolution, no additional explanations are necessary when encountering the full range of human reality. Teleology, mind, free will, and natural theology—any expression of this sort, including even an expression of the idea that all of nature constitutes in an inherent manner a whole, a *cosmos*—is excluded or bypassed. Toulmin's reference to "natural theology" (which is discussed below) was not meant to provide a doctrine about the nature of God with or without reference to the Bible, but rather is an implied reference to these trans-empiric aspects of cosmological research.

Whatever else the philosophy of naturalism may mean, its most identifiable aspect may be the rejection of any religious references or explanatory possibilities. ". . . [naturalist philosophers] would . . . both reject supernatural entities, and allow that science is a possible route (if not necessarily the only one) to important truths about the "human spirit."[12] This antireligious doctrine especially seems aimed at the Christian religion but, also, by association, the traditional Jewish faith and certainly the Muslim faith (given the recent uprising of militant Islam). Yet, the rejection of the three monotheistic faiths, which are also revealed faiths, implies that for the naturalist mind set, certain topics that are only indirectly connected to religious doctrine are also rejected. Teleology, for example, is ferociously repelled by Darwinian "defenders of the truth" although by now, given the complexities that we have now discovered are necessary for the evolution of human life, it is a plausible hypothesis.[13] It is guilt by association, as if teleology is just too close to "natural theology" in Toulmin's meaning, and, taking it a step further, to the theology of the Bible.

Naturalism is a free-floating, vague, but persistent idea that is not easily defended since it would have to be defined in a more explicit manner than has been done thus far by its advocates. Naturalism, however, does refer to science for its validation, but which field of science is appealed to is relevant. The scientific basis of naturalism is usually expressed not in terms of physics because serious discussion of the issues involved, such as the many-worlds interpretation of quantum mechanics (Hugh Everett) immediately leads to philosophical explorations that take place in a space beyond empirical control. Discourses on contemporary cosmology, but also on the environmental consequences of chemical technology and the electronic, computer-enabled invasions of privacy by industry and government, are topics that cannot be usefully debated within the confines of a strict reliance on scientific method.

Naturalism relies, to repeat, on an expanded version of Darwin's theory of evolution, which it doesn't have to defend in any detail, for, in this case "evolution" has become a totem word. Its prior successes in debates with biblical literalists provide Darwinian evolution with its reputation as the all-potent destroyer of religious accounts of creation and of human nature. Because it relies less on argument than social pressure for its acceptance, any full-bore criticism of naturalism is prohibited in the manner of a social *faux pas*, like leaving your date at the bar or burping at the dinner table. How naturalism is less a coherent philosophy than a mode of social integration was on view in the reaction to a recent book by the prominent philosopher, Thomas Nagle. In an analytic, precise, and somewhat deferential manner, he argued against the "materialist, neo-Darwinian conception of nature," which, he stated, "is almost certainly false." His work was treated as if it were a betrayal among the Northeast intellectual class of which he is a part (as a professor at New York University; contributor to the *New York Review of Books*), receiving much harsh criticism.[14]

The Law of Diminishing Reductive Returns

Following the basic pattern of Reductive progress, the three modes of doing science are physical, biological, and social, each of which arrived at its modern form in historical succession. In terms of experimental discovery however, this historical pattern can be understood in terms of the uncovering and discrimination of observable facts, for there is an increasing "degree of difficulty" in the facts themselves, which is the basis for the succession of the three major branches.

Modern science began in the sixteenth century with astronomy, which in epistemic terms is the easiest of the sciences to pursue. Its data points are objects in the sky that are easily discernable to the naked eye and that appear in separate and discrete units, that is, the stars, the moon and the sun. Study of the motion of these "heavenly" objects reveals patterns that were used by the Babylonians and the Greeks to predict eclipses and to roughly determine the circumference of the Earth and its distance from the sun.

Modern scientific biology began with the study of patterns, which are not apparent among the stars but in the internal structures and external forms of living things, and their relations to fossils dug up from the Earth. The fossil evidence implies an enormously large time scale, which, in turn, requires a larger set of different types of data points. Not only is more information required to detect evolutionary patterns, but also different *kinds* of information, such as fossils, comparisons of anatomic forms, distribution patterns of species, techniques of artificial selection, and the like.

As for the social sciences, the relevant data points are not confined to specifiable kinds. Almost anything will do, from the analysis of Grimm's fairy tales (Bettleheim), to the economic statistics of the Great Depression (Keynes), to the willingness of experimental subjects to electrocute a stranger (Milgram). The point is that as the reductive pattern in science history emerges, it reveals that modern science has progressed as limited or defined by the increasing complexity of what it attempts to study.

The epistemological pattern of increasing complexity over time also reflects an ontological progression. In physics and chemistry, the so-called "hard sciences," research focuses on the basic materials of nature, such as atoms and molecules, gases, liquids, and solids. In biology, research examines the qualitatively more complex level that consists of living beings, including cell structure, gross anatomy, ecology, and the study of species as such. This includes exploring their mutual relations, whether as a matter of abstract connections based on bodily characteristics (classification) or historical progression (evolutionary). Finally, the social sciences, in which the level of complexity is at its most intense, tackle the behavior of human beings and animals, in which motive, perception, and intent of living actors must be taken into account.

At each progressive stage, we can discern that some things are left behind as modern science moves forward. These relate to the mythological, philosophical, teleological, religious, literary, and aesthetic connotations/connections. Thus, the modern study of scientific astronomy rejects the ancient connections of star patterns such as the Zodiac, which is based on twelve "houses" that were used to give insight and organization to human personality, and that were and still are the basis of casting astrological charts (something that Kepler and Galileo did at one point in their careers). The association of the classical seven planets, each with their own special characters and assigned godly personages—such as those associated with Venus, Mars, Jupiter, and Mercury—is still evident in popular culture today (for example, in the 1992 book "Men Are from Mars, Women Are from Venus"). Modern scientific astronomy, however, rejects these poetic associations as, at best, pleasant nonsense and, at worst, ignorant superstition; in no case should they be taken seriously as an explanation of planetary motions or of their putative influence on human affairs. On balance, it would seem that educated people may sigh at the loss of astronomical mythology—no more taking the daily horoscope seriously, or using the question "What sign are you?" as an opening at a social occasion, perhaps. But the loss is not likely to be keenly felt or regretted as if it were a tragic loss of innocence.

There are nuances, however. There is a further loss that comes as a result of transforming the starry heavens, the experience of which has always been heavy with implied meanings, as when science reduces them into a map of geometrically defined motions of stars and planets. The stars, pinpoints of bright lights in the inky blackness of the night sky, are referenced in the Old Testament and in the Koran as revealing the glory of God. In a setting beyond city lights, the night reveals a panoply of stars filling the entire sky, a sight that is literally awe-inspiring, giving forth intimations of such things as the smallness of human existence and the glory of God. But, less definitely, there is often present in such observation a sense of simply being in the presence of something universal. Such things as a theory of universal gravitation like Newton's or the Chinese twelve-year system of astrological signs may be equivalent attempts to express the inexpressible. The heavenly sky implies transcendence in some way, and an oversensitivity to the protocols of scientific methodology should not in it itself be a reason for eliminating that sense. Einstein and Kant both found inspiration in the stars in the night sky, one for a sense of universal law, the other for a sense of the reality of God.

The loss caused by the rise of the scientific explanation of the stars is not often felt strongly or commented upon. Such is not the case with scientific biology. In the biological sciences, particularly with regard to the theory of evolution, opposition has been constant and vehement and almost always comes from religious believers. The usual point of contention seems to be that the evolutionary account of the origin of species contradicts the *Genesis* account, in which the origin of each species is the result of a special act of creation by the Lord (*Genesis* 2, 19–20). Further, the *Genesis* account takes place famously in six days, which contradicts forcefully the evolutionary time span in the billions of years. But, as explained by Steven J. Gould in his last book, the rejection of the theory of evolution by religious believers is not really based on its apparent contradiction to the Bible, but rather on the deleterious effects that belief in the evolutionary account have had on popular morality.[15] In the most popular religious book the purpose of which is to attack the evolutionary thesis which is published by the Jehovah's Witnesses, the initial criticism is not based on biblical literalism, but on the decline in morals that resulted from the materialism and rejection of divine revelation that was the evident social effect of the wide acceptance of the theory of evolution.[16] The contradiction between the biblical and evolutionary time spans is easily traversed by an interpretation of the biblical text in which, for example, the days of creation designate not twenty-four-hour days, but multi-million year time spans. What cannot be traversed is the chasm between what is basically a materialistic understanding of human reality and a traditionally religious one. Thus, the objections to evolutionary theory will continue.

The evolutionary example does not include the questionable ethical situations that have recently arisen as a result of the development of new technologies on the cellular and genetic levels. These include cloning and stem-cell research, in which zygotes are treated not as nascent human beings but as raw material for industrial purposes, namely, the mass production of therapies for the treatment of multiple diseases. But if certain essential elements—essential, that is, to understanding human reality in its traditional, humanistic, literary, aesthetic, and religious sense—are arguably rejected in the biological sciences, the situation is most clearly discernible and better defined in the social sciences. Despite the variety of those fields themselves labeled "social science" (psychology, sociology, economics, and the like), one common and essential feature is their treatment of values, whether humanistic or religious. Valuative terms are excluded from consideration in the social sciences, except as objects of study. They are opinions on which to conduct research rather than values to be shared, which is a sharply conceived difference of viewpoint.

Values when shared have a sense of imperativeness about them; in such a circumstance, certain acts ought to be done or, more typically, ought *not* to be done. "Thou shalt not commit adultery!" This attitude is different in surveys about sexual behavior, such as the Kinsey Report. The statistical tables describing the occurrence of certain types of sexual behavior stand there on the page, not as information but as a challenge. If so many do it, the naturalistic urge implies (really a sign of the moral weakness of human nature), can it really be wrong? Who is ultimately to decide what is morally acceptable behavior and what is not? Just as there was in its early days a materialistic ontology underlying physics, so currently there is a naturalistic ethics underlying the social sciences. Moral evaluations resulting in commands, once they are subjected to the kind of analysis characteristic of social science, become denuded of their obligatory force, appearing as only one of an infinite possible variety of reaction patterns.

Understood reductively, the progress of modern science is evident in its successive encounters with the three separate blocks of empirical reality. Naturally, it can be observed, science has progressed from the less to the more difficult, in other words, from physical bodies, to living organisms, to the complex aspects that typify human behavior. But as different blocks of reality have been encountered during the development of empirical science, the difficulties have increased. The sense of loss of fundamental qualities has become more apparent as science has approached more essentially human aspects. This illustrates the law of diminishing explanatory scientific returns—or more colloquially, the further along science progresses, the less it is able to say.

The Philosophic Timeline of Scientific Progress

The Philosophic idea of progress maintains that philosophy of some sort is inevitable from this point forward to understand modern science, for it is no longer plausible that science can stand on its own terms, that is, as intellectually self-sufficient. As its name implies, the philosophic vision of progress, both scientific and social, is distinctive and separate from the previous four. It assumes that science and philosophy are connected and linked such that the main currents of scientific thought, including its major theories, have direct philosophic implications and cannot be understood fully unless a philosophic analysis is applied. This assertion is controversial among defenders of a scientific point of view, and a defense has been made in a previous section by means of examples of how natural scientists themselves have had to do philosophy (and, in one case, theological analysis) in order to make sense of their own fields of research. In terms of an overall time line, the philosophic model assumes, as do all the prior models, that there was a "revolution" or a vast separation between what can be usefully termed the "medieval world-view" and the modern, scientific one that we have inhabited for the past four centuries. It is argued that we are now in a phase of transition in which modern science becomes again entangled with philosophical origins.

The transition can be seen in the manner in which historian Paul Johnson perceived the development of evolutionary psychology as a new scientific advance. The advance established that, on a basic biological level, human nature is a constant, not a variable, thing that is based on social construction and that thus works against what, in Johnson's view, is the "relativism" implied by Einstein's eponymous theory.[17] That transition can be further understood and described by taking an account, however brief, of current cosmological speculations. However directly based on empirical data and quantitatively described they are, these speculations cannot "escape" certain philosophic implications concerning the existence of the human species, which can no longer be plausibly understood as a trivial accident of cosmic evolutionary processes. One of the problematic issues raised by the new cosmology is its intense level of mathematical abstraction, which seems to imply that scientific reality is in essence nonmaterial and exists as a Platonic form. In both thought and form, all of these new developments have a particular resonance in contemporary society, in this postmodern era of anxiety, regret, and attempted restoration/resolution. And the philosophical understanding of scientific progress has perceptible implications for how science itself is to be understood in terms of the relationship between the lived reality of human life and the nature of scientific abstraction, and subsequently in terms of the current state of postmodern culture, which is far less optimistic than it was a century ago.

Taking the four prior visions of progress in turn, we can compare them to the newly emerging philosophic vision of scientific progress. Compared to Whig progress, the philosophic version likewise assumes that science does make progress in accumulating data from empirical observation and experiment, and as well in the increasing sophistication, comprehension, and explanatory power of its theories. The philosophic concept is, however, less convinced that a final, grand theory that unifies all science is on the horizon, even in the long term. The Philosophic version can only make such claims and criticisms if the concepts and methods of modern science are connected in an intrinsic manner to Philosophic reasoning and concepts.

Compared to Enlightenment progress, the Philosophic version also looks to the origins of scientific thought in ancient times. But it does not see medieval times as an interregnum in the development of scientific thought; it accepts in some degree the "continuity thesis" that medieval thought not only preceded but enabled the arrival of modern science. Philosophic progress sees Bacon's philosophy as a caricature of Aristotelian philosophy and an unfair condemnation of both ancient and medieval thought. Inductive reasoning was well-understood by Aristotle, who had a more sophisticated understanding of the interplay between and mutual dependence of inductive and deductive reasoning than did Bacon, as in his *Prior Analytics*.

Compared to Reductive progress, Philosophic progress opposes the claim that scientific explanation is able to reduce philosophic (as well as political, religious, artistic, and historical) insights perfectly to empirical facts and mechanical explanations, with nothing left over, so to speak, except useless regret. It is not enough to merely oppose reduction, for such a stance is not unusual, as Weinberg reluctantly admitted. The claims of the Philosophic version of progress must contend with the plausible description of the Reductive model. This model is based on the idea that as science has progressed, it has done so by eliminating in a progressive and step-by-step manner all teleological, mystical, intuitive, traditional, religious, and philosophic descriptions and explanations from each specific area of scientific interest: astrology eliminated from modern astronomy, teleology eliminated from modern biology, and moral consciousness eliminated from modern psychology. The philosophic model however, accepts the "law" of diminishing returns, described above, that as modern science has progressed into increasingly complex and difficult areas of research, its methodology and mechanical concepts have had less to say. While the removal of astrological myth from astronomy seems to result in no loss, the completely mechanical description of the heavens also eliminates the sense of awe and wonder that leads the observer of the night skies to a beneficial humility, as experienced by the Psalmist, Mohammed, Kant, and Einstein.

There is a closer resemblance between the Philosophic and the Historicist versions of progress than among the other models of the history of science. As with Historicist progress, Philosophic progress assumes that there is not a very clear demarcation between scientific statements and concepts and those derived from philosophy and the general ambient culture. Like Historicist history, Philosophic history also sees that various chapters of scientific thought come not as single discoveries, but as a collection of theories, tested observations, research methods, instrumentation, and mathematical techniques (as in, for example calculus, probability, vectors, tensors, matrices, and the like) that form a basis not only of intellectual but also of social identity. In other words, science is understood as a human enterprise that is a means of denying the triumphalist claims of reductive science, and also a means of understanding modern science in a human and humane manner, that is, in a historical context. There is a general difference, however, for Philosophic history rests on conceptual analyses that contradict the implicit irrationalism of the Historicist view, that results in an inability to make comparisons or to generalize about the progress of modern science.

Wittgenstein, Toulmin, and Natural Theology

Ludwig Wittgenstein was probably the most important philosopher of the English-speaking world in the twentieth century. He did not propose any new synthetic view of knowledge or new system of metaphysics, scientific method, or practical ethics; rather he proposed that the traditional philosophic arguments and problems were the result of confusion in the manner in which philosophers misused language. This seemingly limited view of what philosophy could accomplish nonetheless had a great influence on his hearers and disciples, such as G.E.M. Anscombe and Norman Malcolm. They viewed Wittgenstein's influence as a form of release from ideas that were assumed as part of the intellectual ambiance of the times, including the empiricist account of knowledge and the implicit materialism and explicit positivism that were dominant in the intellectual culture of the day. This effect was the deliberate result of Wittgenstein's methodology, which proposed that philosophy—and basically all intellectual enterprises—ought not to be constricted by what Wittgenstein claimed were the noxious results of the misuse of language. He believed that, in philosophy, such misuse hardened questions and answers into intellectual knots that could not be untied. His philosophy, as he noted, is a kind of language therapy for philosophers. A number of Wittgenstein's followers including Anscombe, were religious believers.

Among Wittgenstein's disciples was Stephen Toulmin, who was not a religious believer as far as is known but whose writings on modern science and cosmology included the notion that "natural theology" was an inherent component. Toulmin had experienced the same rush of intellectual freedom that other Wittgenstein disciples did, a development that is apparent in Toulmin's writings on the philosophy of science, particularly cosmology. Toulmin approached these areas in a manner quite different from that of his contemporaries, including Carnap and Popper. He was prepared to notice general developments that lay beyond the limits of empirical experiment and exact logical protocols, but that were evident in scientific theorizing nonetheless. Among the topics approached by Toulmin was the topic of cosmology.

As Toulmin points out at the outset, cosmology is the study of the whole, "thinking about the universe as a whole."[18] From considerations of the history of ideas, it is apparent that such speculation arose in ancient times and has continued until our own day. But, necessarily, consideration of the whole of existence cannot be *a priori* limited to one approach or another. Thus, it is illegitimate to confine such speculation to the discoveries of empirical science alone, as atomists and scientific materialists have done. And in any case, there is a hubristic naïvete in assuming that the human intellect can approach such a vastness of knowledge without an understanding that scientists are a part of the nature they are researching, and of the innate complexities of the process of knowing itself (Kant). Ineluctably, then, cosmological concepts—no matter how dependent on and derived from scientific theories—will contain philosophical aspects.

The first of two books he wrote on the topic, *The Return to Cosmology: Postmodern Science and the Theology of Nature* was essentially a collection of several of Toulmin's prior shorter writings, including extended book reviews It was wrapped coherently around the single assertion that recent developments not only showed a renewed interest among scientists in explaining the universe as a whole, but that this development in scientific thought also implied that the subject of *natural theology* was a proper concern in cosmology. This was considered in the sense that speculation about divinity and its relation to physical nature and the nature of human knowledge was now a permitted aspect of an expanded idea of nature and natural science. In effect, this possibility of natural theology was not a radical imposition of assertive theology onto the turf properly limited to scientists. It was, as the title of Toulmin's essay indicated, a return to what had been the common interest of natural science in its beginnings. It is relevant here to note that the title of Newton's profoundly influential work is most often presented with a truncation of its full title, which deforms its purpose. It is almost always referred to as the "Principia Mathematica," or even "the Principia," when its true title is "Principia Mathematica Philosophiae Naturalis"; Toulmin repeated the title in full. The Latin is the same as the English, for the nouns share the identical meaning; what Newton thought he was writing was a *philosophy of nature* according to mathematical principles, and thus he did not limit his theorizing to the direct empirical evidence that supported his great theory. How could it be so limited when Newton had to reach speculative heights to extend his reasoning from the fall of an apple to the revolutions of the moon, and when his general theory of gravity was proposed to be effective on every physical body throughout the whole universe?[19]

Toulmin's approach in the third and final section of *Return to Cosmology* is more literary and historical than scientific. Toulmin deals with the separation point between the medieval world-view and the modern in the chapter "All Cohaerance [sic] Gone," referencing an older spelling as well as Donne's poetry and Plato's *Timaeus*. But he makes the relevant point when, at the beginning of that chapter, he proposes to link "postmodern science and natural religion," which he is aware immediately raises serious questions. One such example is: "Surely, the fates of natural theology and natural religion were settled long ago; so how can we discuss them constructively today?"[20] He defends his countercultural proposal by reference to Newton. Until 150 years ago, these questions would scarcely have seemed so puzzling. Isaac Newton, that devout Protestant mathematician, did not hesitate to discuss the actions and attributes of God, the Creator and sustainer of nature, in the context of his own scientific writings. For instance, in the "General Scholium" that he added to the second edition of his classic *Principia Mathematica Philosophiae Naturalis*, Newton insisted that "to discourse of God does certainly belong to natural philosophy." Toulmin does not support his argument about natural theology and modern cosmology with evidence provided by Newton's example alone; the preceding portion of his book consists of a selection of essays (previously published in *Encounter* and in the *New York Review of Books*) on recent scientific figures who have, in effect, erected a cosmology containing definitely philosophic aspects. Among them: Arthur Koestler, Teilhard de Chardin, Jacques Monod, Carl Sagan, and Gregory Bateson.

Toulmin, however, does not specify what "natural theology" actually consists of. That is, he offers no content but is satisfied to merely make the point that such theologizing was a part of scientific cosmology up to 150 years ago, and that in the late twentieth century, is making a comeback, albeit in some indefinite but observable manner. In other words, it is not really a theology that most of the writers he reviews are doing, except for Teilhard, of course; Bateson and Monod explicitly deny the existence or the intellectual functionality of the concept of God as it is traditionally understood, for example, as expressed in the Nicene Creed. Toulmin refers to *natural* theology, which seems to exclude the God, or Lord, that is revealed in the Bible. Rather another concept of divinity is implied, one that is circumscribed by the limits of human reason alone (Kant). What Toulmin is actually referring to is that in postmodern science, we are involved in "reinserting humanity into the world of nature," for . . . "scientists have always to consider themselves as agents, not merely observers, and ask about the moral significance of the actions that comprise even the very doing of science."[21] (The intense ethical issues surrounding the creation of the atom bomb by the scientists at Los Alamos come to mind.) As he analyzes the recent interest in cosmology, however, he detects resistance. Toulmin points out the effect of the professionalization of learning and the division of human knowledge into separate and competing university departments. The unity of knowledge is surely lost in the mechanization not just of the world picture but of university organization.[22]

Notes

1. New York: Pantheon, 2014.
2. Lawrence Krauss, *A Universe from Nothing: Why There is Something Rather than Nothing* (New York: Free Press, 2012.
3. See *Exodus*, Chp. 17, vs. 12.
4. Galileo Galileo, "Letter to the Grand Duchess Christina" in Maurice A. Finocchiaro, *The Galileo Affair: A Documentary History* (Bekeley, CA: University of California Press, 1989), 87–118.

5. http://plato.stanford.edu/entries/qt-epr/.
6. Rollo May, ed., *Existential Psychology* (New York: Random House, 1961), 88.
7. "Some Issues Concerning the Control of Human Behavior", *Science*, Nov. 30, 1956, Vol. 124, No. 3231, http://www.sciencemag.org/content/124/3231/1057.extract; .1057.
8. Burris F. Skinner, *Beyond Freedom and Dignity* (Indianapolis, Hackett, 1971).
9. Abner Shimony, *The Search for a Naturalistic Worldview* (Princeton; Princeton University Press 1992).
10. For essays on Shimony, see *Potentiality, Entanglement and Passion-at-a-Distance*, ed. by R.S. Cohen, M. Horne, J. Stachel (Dordrecht, Kluwer, 2010).
11. http://plato.stanford.edu/entries/naturalism/
12. Ibid., accessed 6/14/2014.
13. John Caiazza, *The Ethics of Cosmology* (New Brunswick, NJ: Transaction, 2012), xix.
14. Thomas Nagel, *Mind and Cosmos* (New York: Oxford University Press, 2012). For reactions, see http://www.themontrealreview.com/2009/Darwin-and-Dogma.php.
15. Watch Tower Bible and Tract Society ("Jehovah's Witnesses"), *Did Man Get Here by Evolution or by Creation*? (New York, 1967), 130.
16. Stephen J. Gould, *Rocks of Ages.*
17. Paul Johnson, *Modern Times* (New York: Harper and Row, 1983), 733–34.
18. Stephen Toulmin, *The Return to Cosmology*; 1, 2.
19. Ibid, 371.
20. Ibid, 217.
21. Ibid., 255–56.
22. Ibid, 229–31.

8

Cosmology and Human Existence

Toulmin writes that scientists from now on will face the unavoidable task of understanding themselves as moral agents. They will no longer be able, as it were, to take a view of their study of nature from the outside or in an objective fashion, as if they were observing a play rather than taking part in the action. Thus, they must focus on the manner in which they do their research as well as its theoretical conclusions. But Toulmin's argument, which has been accepted here, is still an argument made in general. To be more specific, there are several areas in which recent, postmodern science manifests the "reinsertion of humanity." The first area refers to cosmology, namely, in the role of the observer in relativity and quantum mechanics; the second concerns acknowledging the insistent presence of human existence as a "phenomenon" within modern cosmology. But the reinsertion of humanity has further implications for the application of biological and neurological fields in the study of human nature. These directly imply that there is such a thing as human nature, common to all members of our species. This conclusion contradicts the progressive view in a direct fashion: if human nature is permanent, has in effect an "essence," then progress based on the elimination of certain characteristics involving gender, social hierarchy, or aggressiveness cannot take place beyond a certain point or, if so, only at an immense social cost. Finally, prominent philosophers who currently concern themselves with social ideals no longer directly depend on the social sciences to lead them. Instead, they prefer to refer their thought to ancient and medieval models, especially Aristotle's philosophy, which to them seems more comprehensive and frankly realistic; sociological theories and psychological models are seemingly of not much use to understand and prescribe social ethics for human society.

Cosmic Role of the Observer in Postmodern Physics

"Relativity" in Einstein's theory does not mean that all things are "relative," including ethical systems; that idea is an illegitimate extension that implies the relativity of morals. In the context of physical theory, "relativity" means that if two different dynamical systems are in place to describe the same set of phenomena, and both are mathematically consistent according to the laws of classical mechanics, then either system can be chosen without further consequence. This applies to the Ptolemaic and Copernican descriptions of the solar/planetary system. "Relativity" means that one system can be transformed into the other system and vice versa by following a set of transformation rules. Therefore, the accelerated motion in one system can substitute for the effect of gravity in the other, and energy can be transformed into mass and vice versa according to the iconic equation, energy equals mass times the speed of light squared. But the transformability of Einsteinian relativity does imply that the stance observers take, whether in place on the surface of the moon or traveling at light speed on a ray of light, affects what they observe and how they describe it mathematically; there exists no preferred place within the cosmos from which a universal "take" is possible. Hence, the emphasis is on what philosophers of science have termed "the role of the observer."

The issue is apparent in quantum physics as well, so well-known in fact that only a reference to the issue is required at this point. In the EPR thought experiment that is explained in an earlier chapter, the effect of the observation of one entangled particle immediately affecting the other, necessarily requires for its completion that the first particle be observed; hence, the inevitable discussion of the role of the observer in quantum mechanics. What is of interest here, however, is the degree to which physicists in general have had to deal with or else ignore the philosophical consequences of quantum observation and quantum mechanics in general. Immediately after the discoveries and major theories that came to define the radical new field, scientists were aware of the philosophic issues, particularly the indeterminism of subatomic processes. Einstein opposed the inherently probabilistic nature of quantum mechanical explanation and the inferences made about the nature of subatomic reality. Today, however, the philosophic issue has moved on; there have been no times of crisis in the last twenty or thirty years, and the field of quantum physics has been happily applied to provide numerous technological advances in electronic communication and computer chips. As a result there is no longer the intensity of philosophic awareness apparent in the field, as there was in the olden days when scientific giants roamed international conferences at places like Solvay.

Another reason for the lack of philosophic interest in the field and in the question of quantum observation is that quantum mechanics has developed its own ersatz form of philosophy, which is often referred to as the idea of "The Many Worlds Interpretation," invented by the otherwise obscure figure of physicist Hugh Everett.[1] Given that in any quantum observation, two experimental results are possible and, further, that only one of them will actually take place, the issue arises, what of the other possibility, the quantum road not taken? What is its status, and how can one avoid the Berkeleyan consequences of quantum observation, that is, that the observers/experimenters are, in effect, creating the thing they are observing? Einstein had noted the problem, referring to the iconic Latin term, *esse est percipi* ("to be is to be perceived") of the Idealist philosopher and Anglican bishop Berkeley.

Everett's proposal, accepted by many in the field it seems, is that in a quantum observation both of the two possible results take place, but that the one that we cannot detect exists as an alternative history in a universe or a state separate from the one we inhabit. In this way, the issue of whether quantum phenomena are self-created is avoided, since both possible results of a particular act of observation or experimentation are realized. The intellectual effrontery of this solution is sometimes hard to grasp; to the outsider, whether a philosopher or an educated person with a general interest in science, it seems to violate common sense because it means that we must envision the reality of an alternate history that comes into existence at every point at which a quantum observation is made. A cosmic development of the alternate history account comes about when it is applied to the origin of the universe, for the big bang account starts with a subatomic particle, the ultimate subatomic particle from which all physical existence began. Applied to cosmic origins, the alternative history theory implies that other whole universes besides the one that astronomers study and that the human species inhabits exist, in some manner, besides our own. Everett's alternate history theory is now swollen into an alternate universe account that implies not one but a multitude of possible alternate universes.

As the philosophic observer contemplates such exotic speculations, including not just alternate universes but additional dimensions (up to eleven) and string theory (another ultimate subatomic "reality"), one wonders whatever happened to the chief brag from advocates of science that its method is superior to the speculations of philosophers and the circular certainties of theologians? Whatever happened to scientific skepticism, that has been applied with such reductive intensity to the traditional ideas of free will, immaterial souls, and teleological explanations? And whatever happened to Ockham's razor, the principle that entities and explanations should not be multiplied without necessity? And, finally, whatever happened to the scientific ideal that its theories are based on positive, empirical evidence, for as a matter of principle it seems that the existence of alternate histories and alternate universes are beyond empirical observation? And, yet, these exotic speculations are touted as the last word in science's explanation of all things in the universe and beyond, promoted like a commercial product complete with best- selling books and glossy documentaries on public television.

If there is necessity for such speculations, it derives from the inability of mathematical physics, despite its recent successes, to be able to offer a comprehensive theory of everything. Also, the outside observer expects, there is the fear among the chief scientific theorists themselves that the theoretical difficulties they are currently grappling with are not based on a lack of experimental validation but derive from some ultimate limits of scientific methodology itself. But if that is the case, then speculations about alternate histories, alternate universes, and the like become the predictable consequence of explanatory crisis in physics at its highest level. It is surely useless at this point to claim that scientists are not doing philosophy, as they are in fact encountering issues and problems that philosophy has been treating for millennia. Not only Berkeley has been referenced as a matter of course in recent speculations, but Plato has as well, for it seems that if such alternate realities exist that they do so in a purely abstract and mathematical space.

Scientific Cosmology and Human Existence

(Please note: this section originally appeared in a different form as a paper prepared for the conference "The Value of Human Life," held at the Franciscan University of Steubenville's Institute of Bioethics. It was subsequently republished in the online journal *InSight*, edited by V. Ryabov and published by Rivier University.)

By one definition, cosmology is that branch of philosophy (or science) that "concerns itself with the origin and general structure of the universe, its parts, elements and laws, especially with such characteristics as space, time, causality [and] freedom." Up until recently, such a comprehensive look at the universe has been thought to be the province of philosophy—and speculative philosophy at that. Now, however, cosmology has been developed as a part of empirical science mostly as a result of Einstein's Theory of Relativity which, with mathematical precision, enables science to describe the point of origin of the universe and its general structure, including space, time, and gravity. On an imaginative basis, what Einstein did was to enable us to conceive of the universe as an optional thing; in other words, to imagine that the universe may have been other than it is. Alternative universes may be imagined to have too powerful or too weak a pull of the subatomic force that binds atoms and molecules together for living matter to form. We could have had a universe without planets or stars, simply hot gases with no solid matter, or, alternately, a universe that would never attain a temperature level much above absolute zero. Such universes obviously could not support life, which in fact is true of most universes that may be imagined scientifically. As for *this* universe, the one that we know actually exists, as scientists continue to pursue its fine-grained parts and processes they have found an increasing degree of complexity, and some have concluded that the present arrangement of our universe is too complex to have happened by accident—if, that is, life is to exist. The underlying issue in the development of scientific cosmology is that this universe supports life, and especially intelligent life.

There is an attitude discernible among certain scientists and some philosophers and popularizers that dismisses or deprecates the value of human life. That attitude is encapsulated in what is called the Copernican Principle, which is described as "one of the primary pillars of the science of astronomy [which] says simply that we do not occupy a privileged location in the universe." Just as Copernicus removed the earth, mankind's home, from the center of the solar system to replace it with the sun, so in general it is claimed that one of the aims of science is to displace mankind from any significant or unique place in the universe. The Copernican Principle is, according to this view, verified by Darwin's Theory of Evolution because it reduced mankind from a special place in nature, ordained by God to be the apex of creation to merely another species of animal, albeit the most advanced one. The Copernican Principle, it is fair to say, has been elevated from a rule of thumb that is useful to assess the relation between human observation and astronomical phenomena to a philosophical principle that asserts the unimportance of human existence from a scientific point of view.

There is a counter to this denigration of the value of human life that comes from scientific cosmology. The revolution in twentieth-century physics can be described in general terms as the replacement of the mechanical view of the universe, which was essentially static, with a relativistic view, which is dynamic. The older view was based on the mechanical science of Newton in which the universe was pictured as a vast machine: infinitely large but mostly empty, containing only the occasional star, planetary system, or galaxy the motions of which were completely predictable by mathematical laws. Pascal said that "the silence of those infinite spaces terrifies me," since the Newtonian mechanical universe was precisely indifferent to the existence, purposes, beliefs, or strivings of human beings. The pre-relativistic version of the universe was also indifferent to the subject of cosmology, and thus pushed the subject of the universe as a whole almost entirely into the realm of philosophic speculation. But then, there was not much that was interesting about the mechanical vision of the universe; there was no mystery and no engagement of the sense of awe or beauty in a something that was essentially a vast, empty space with nothing to differentiate one part from another, and from which any possibility of meaning had been eliminated. Even time was somehow absent, for, odd as it may seem, the passage of time of which our culture has become almost pathologically aware did not affect the atoms and forces that constituted the mechanical universe. Time was an axis along which the mathematical description of motion moved, but in theory the time coordinate was completely reversible. It made no difference whether the physical processes were played forward or backward; the outcome was always the same. From a beginning point, Laplace stated that given enough information, the paths of particles and bodies could be completely predicted; from the condition of an endpoint, the beginning position and momentum of the particles could be accurately determined. But this vision was subsequently scrapped in favor of the evolving, dynamic version of the relativistic universe that is now favored by modern science.

The relativistic universe has a *lifecycle* like a mayfly or a man, and, as is often said, is undergoing *evolutionary development*. These metaphors are not taken from machines but from life, so the new model of the universe is not only time- bound, it is in a sense organic. The new cosmology has not only uncovered the fact that universe is dynamic; it has also made discoveries about the nature of matter on the atomic, molecular, and subatomic levels that reveal the fine details of physical reality that make life, and human life in particular, possible. A whole host of physical facts and constants have to be precisely what they are in order for life to exist, and such facts can be discovered at all levels of physical reality. On the atomic level, the peculiar inner structure of the carbon atom enables it to form chains with six atoms linked together. From each of the six carbon atoms, a string of other atoms and molecules can extend, allowing the existence of the complex organic compounds that are the physical basis of life. On the astronomical level, the position of the Earth is not too close to the sun like Venus, or too far away, like Jupiter, and so is in just the right orbit for life to evolve and exist. On the middle level, water, which was praised by St. Francis as "very useful, humble, precious and pure," has unique qualities among chemical compounds that are necessary for the existence of life. Due to its peculiar chemical structure, for example, water in its solid state, as ice, is lighter than its liquid state. This means that lakes and rivers do not freeze from the bottom up but from the top down, enabling aquatic plants, fish, and other life forms to survive over the winter. As science has proceeded to uncover the complexity of the physical universe in ever more detail, the conditions necessary for life have added up until the probability that life arrived in the universe and on the earth *by accident* is no longer plausible.[2]

Added to these newly acquired facts is a new appreciation of what has always been true about scientific discovery, namely, that it is human beings with their wills and intellects who make the controlled observations and who devise the abstract theories that are the essence of scientific inquiry. The fact that human beings have evolved as part of the universe, and that they may be discerned to be an intended part of it and that their existence is necessarily involved with the detection of scientific truth, is encapsulated in a new principle, the "Anthropic Principle." This principle asserts that the universe cannot be understood without including the fact of human existence, that is, to use Teilhard's terminology, science has to take into account "the phenomenon of man." Much has been written about this principle and whether it has a place in physics. But it seems plausible that we may take the anthropic principle as contrary to the Copernican Principle, though both have the same status as a metascientific or metaphysical idea, which current physics finds useful.

The Anthropic Principle has several variations, from "strong" to "weak." The strong version asserts generally that the universe evolved for the purpose of human intelligence and, further, that it would not exist if it were not for the presence of human intelligence. The strong version is not accepted by most scientists, who have considered the matter and feel it remains highly speculative. However, the weaker version of the anthropic principle does have a number of respected advocates. One weaker version asserts that in studying the cosmos as a whole, the fact of human existence makes some physical values more probable than others. The physical properties of the universe have to allow for the presence of sites where *carbon-based life forms* can live, and, further, the universe must be old enough for life to have evolved. This might seem obvious on its face to a nonscientist, but the weak version still faces significant resistance. One of the advocates of the weak anthropic principle is Steven Weinberg, a Nobel Prize-winning particle physicist and a renowned author on scientific topics for lay people (*The First Three Minutes*). But because Weinberg has attacked philosophy as a means of knowledge in favor of scientific method, he has not accepted the weak anthropic principle for philosophical or quasi-religious reasons. However, because as a physicist as he considers cosmological alternatives, he has deduced that there must be a boundary that allows for the existence of the one universe that we humans know as an existential fact to exist.

Human life may now be seen to be built into the fabric of the physical universe, and its value can have an ultimate moral and intellectual basis. Otherwise, asserting that human life has a value may be seen as an imposition on an otherwise uncaring universe that is unresponsive to human needs and not specifically intended for human existence. The discussion about the Anthropic Principle is an indication that human existence can now be accepted by the hard sciences as part of the "fabric of the universe" and not an accidental or adventitious thing. As heartening as it is that some hard scientists have come around to recognizing that human beings are an intrinsic part of the cosmos, we should be aware that it is a limited concession. The part of human nature they are concerned about is human intelligence; the moral aspects of human nature are not part of their equations. On the other hand, there is no denying that there are religious implications underlying the discussion. If human life is somehow purposed, is there a maximum agent who acts on the purpose?

Ten Crucial Steps in the Evolution of Man

Each of these steps was essential if the human species was to develop. The arrival of each one was improbably singular, and the combined improbability of all the steps makes the arrival of human life a statistical impossibility. (Taken from *The Anthropic Cosmological Principle*, pp. 562–564.)

1. The DNA-based genetic code is universal among living things, and is the basis for reproduction.
2. Aerobic respiration, which is essential for life in an oxygen atmosphere, especially for multi-celled life forms.

3. Glucose fermentation, which is essential for energy production in multi-celled organisms; it is complex and requires control by many, not just a single, genes.
4. Photosynthesis, which enables plants and bacteria to generate oxygen from CO2, and thus generate the oxygen atmosphere for multi-celled organisms.
5. Mitochondria in cells, which synthesizes the energy molecule APT, without which multi-celled organisms could not function.
6. Nerve cells, which develop during the division of the chromosomes in the course of reproduction as the "spindles" or bundles between them.
7. The basic eye precursor. Eyes appear several times in the evolution of life forms, but the there is likely a basic genetic pattern behind each separate appearance.
8. The internal skeleton that is necessary for support of large terrestrial animals.
9. Chordate species that depend upon the development of the backbone essential for support of a complex central nervous system.
10. *Homo sapiens,* whose evolution was made unlikely by its high level of intelligence, which is a barrier rather than an aid to survival in primitive conditions.

To these steps must be added two possible interventions from outer space: complex bio-chemicals such as proteins or amino acids arriving on the primitive earth by means of an asteroid which began the evolution of life itself; and the comet whose explosion covered the earth's surface with dust for two years resulting in the death of the large dinosaur species, making room for the expansion of mammalian life forms.

Two Concepts of God: Scientific and Religious

As modern science now considers the big questions, it is proper to think that there is not a "God delusion" but, rather a "God dilemma"—or even two such dilemmas. That is apparent in much recent "God talk" among physicists, including those such as Hawking who go out of their way to deny the existence of God. Carl Sagan states, in the introduction to Hawking's best seller, *A Brief History of Time*, "This is also a book about God . . . or perhaps about the absence of God. The word God fills these pages. Hawking embarks on a quest to answer Einstein's famous question about whether God had any choice in creating the universe. Hawking is attempting, as he explicitly states, to understand the mind of God."[3]

However, Hawking finds no evidence of God as creator in his popular account of the astrophysical universe "from the big bang to black holes"; in his later book he is even more explicit that a creator is an unnecessary means of understanding either the origin or the arrangements of the space-time universe.[4]

Dawkins's *The God Delusion* is the point text here, as it is an attempt to settle the question once and for all. Much debate and criticism has been provoked by Dawkins's famous, or infamous, book, but I will only note here that despite his intentions, the book does not settle the question about the validity of religious belief (i.e., whether it is in fact a "delusion" or whether God actually exists). For Dawkins must in the end rely on a materialist version of the evolutionary narrative to fill in the gaps that God once filled to explain human reality and cosmic history. Frankly, it seems really doubtful that, once the details are considered, evolutionary theory can actually do this. But a historical account of the recent controversies, while not decisive with regard to the existence—or nonexistence—of God, does indicate that a soft kind of crisis is occurring in contemporary scientific and meta-scientific thinking. That crisis can be explained in terms of not one but two dilemmas.

"Einstein's famous question" as Sagan calls it, of whether God had a choice in creating the universe as he did, also refers to "the mind of God." The famous term has been used subsequently by theoretical physicists, including Hawking, to attempt to explain to themselves and to others what it is they are ultimately pursuing in their research work. But why God, and his mind in particular? Einstein was explicit that he did not believe in God in the religious sense, but rather accepted a version of God that Spinoza had described as the expression of rational world order, a version that theologians and philosophers have classified as "pantheism." What Einstein, and other theoretical physicists are referring to here, then, is not "God" in the sense that is traditional in Western culture, but something else. But definitely not a *someone* else, for God in this version is not a person, and this differentiates what we may roughly call the "scientific" view of the nature of God from the religious or biblical view in which God is most definitely a person. How can we further describe the two different versions? We can further contrast them by calling them *God-sci* and *God-bib* by means of descriptions from two of their foremost representatives. An exploration by Einstein of God-sci, as quoted by Lincoln Barnett, states: "My religion consists of a humble admiration of the illimitable superior spirit who reveals himself in the slight details we are able to perceive with our frail and feeble minds. That deeply emotional conviction of the presence of a superior reasoning power which is revealed in the incomprehensible universe, form my idea of God."[5] Einstein here exhibits a humility not shown by his successors, including Hawking and Weinberg, by declaring that the universe will in effect remain mysterious to us in parts or more likely as a whole. He also has a seemingly mystical response to the experience of attempting to understand the universe, which he writes about in other places later on. But in this passage Einstein effectively reduces that mystical insight to a feeling, a matter of emotion. So God-sci for Einstein becomes an evanescent concept, a tribute really to the attempts by scientific minds to comprehend the universe as a whole.

By contrast, Cardinal Newman, perhaps the foremost Christian theologian of the nineteenth century (and, like Dawkins, a leading scholar at Oxford) supports God-bib. After distinguishing the concept of God from a world principle (see Spinoza), a beginning cause (see the big bang), and a projection of humanity (see Feuerbach), Newman describes the God of the religious believer in high Victorian prose.

> I speak then of the God of the Theist and the Christian. A God who is numerically One, who is Personal . . . the life of Law and Order, the Moral Governor; One who is supreme and Sole; . . . unlike all things besides Himself which are all but His creatures; distinct from, independent of them . . . One who is self-existing, absolutely infinite, who has ever been and ever will be . . . who is all perfection. . . . One who is All-powerful, All-knowing, Omnipresent, Incomprehensible . . .[6]

There are two dilemmas here. The first the concerns which version of God to accept when the question of the existence of God arises, as it does in the context of recent cosmology and in the consideration of the mathematical improbability of the evolution of human life. There are certain lines of research that seem to tend this way. One of these is cosmology, which is the consideration of the universe as a whole. This area of research exhibits aspects of unity and is soaked through with anthropic rationality, evoking (as Einstein said repeatedly) awe and a sense of mystery. And from the standpoint of explaining the origin of intelligent, that is, human life, even as he attempts to discount it Dawkins explains what is in effect a new version of a proof of the existence of God, which is assumed to be God-bib, namely, the "argument from improbability." For it seems literally unbelievable that given the degree of complexity in terms of a certain number of events that had to have taken place over time in order for life to evolve, especially intelligent life, that it is all the result of an accident. This new proof, which is a high-tech version of the cosmological proof, concludes that there must be a purpose inherent in the successive operations of the universe for life, and for human life, to have evolved; it is convincing proof for many educated people who have considered it.[7] Thus, this is the first dilemma, whether or not to accede to the multiple strands from cosmology and the improbability of the evolution of human life that cause a kind of attraction, an intellectual gravity field, which implies either cosmic design or the existence of God as a cosmic mind, specifically, God-sci.

But here arises the second dilemma, that is, if God-sci is accepted or even hypothetically considered, how do we keep that concept separate from God-bib? In ancient times, before the arrival and impenetration of European culture by the Christian religion, the concept of God as per Aristotle and the Stoics, for example, could be more easily distinguished from the religious beliefs of the hoi polloi. The general populace understood divinity in a polytheistic sense, as comprising a panoply of very human gods who were representatives of natural forces like the weather or the sea, or interpersonal forces like love or motherhood—beliefs that could not sustain the critical analysis of philosophers such as Xenophanes. Aristotle's version, his famous "unmoved mover," was too philosophical to ever be confused with the beliefs of the common people. Not so in a Christian or even a post-Christian culture, in which confusion resulting from the inability to distinguish God-sci from God-bib seems inevitable.

There is an inevitable leakage between God-sci and God-bib in the common imagination; as a result, accepting God-sci seems automatically to imply acceptance of God-bib. In order to reject this identification, scientifically minded people have had to enter into realms of theology, to which they are unaccustomed. Defenders of scientific naturalism generally, as well as scientific atheists who wish to oppose the possibility of God-bib find, themselves having to oppose it constantly as it keeps rising up in the alternative form of God-sci. This is evident in their attacks on arguments based on probability, the Anthropic Principle, or the need for an ultimate quantum observer. The scientific atheists must play a kind of intellectual whack-a-mole, attacking God-sci whenever it appears among the nonmaterialistic scientists and philosophers who, as a result of recent developments in physical cosmology, find divine speculation impossible to resist.

Recent Science Reveals the Permanence of Natural Human Differences

One of the primary assumptions of progressivism is the malleability of human nature, such that, by means of education and (if necessary) direct governmental policies, immemorial prejudices and habits of social behavior can be changed, even radically, to fit the mold of what a progressive society ought to be. Persons of an advanced age in the United States will remember vividly how the Civil Rights movement, Jim Crow laws, and especially overt expressions of racial prejudice were effectively eliminated within a generation. This was accomplished by a series of famous Supreme Court decisions, but also by forceful use of federal power, as when Eisenhower used the National Guard to integrate the high school in Little Rock, Arkansas. Racial prejudice was also countered by southern journalists who wrote appreciatively about the struggles of southern blacks, and when black culture began to extend into the American consciousness beyond Gospel music and jazz. Indeed, it was the success of the Civil Rights movement that gave inspiration to subsequent movements such as the women's movement and now gay rights. In these profound social changes, in provoking them and in analyzing them, social science was of paramount importance; the *Brown v. Board of Education* decision referenced social science research from the segregated schools of the South to prove the negative result of segregated schooling on black children, for example. Resistance to social change has been explained by the presence of "status anxiety" and the existence of an "authoritarian personality" type. All this fits into the Whig version of progress, and its variations.

In the last thirty years or so a contrarian aspect of contemporary science has presented a challenge to Whig progressivism, for it has become increasingly apparent that human nature is not as malleable as progressives have assumed. Research from biology, neurology, sociology, economics, and political science now strongly indicates that human nature and human society have certain permanent structures that exist beyond the range of the control of progressive policies and explanations. As the results of such research have become known, the pressure on the progressive mind-set has increased, resulting in a series of conflicts—both intellectual and political—that in some cases have culminated in violence on college campuses and in lecture halls. The researchers who presented the new scientific discoveries, which indicated that human nature is a constant thing that cannot be manipulated beyond certain points, were in the 1970s and until recently shouted down, harassed, and vilified. The founding figure of sociobiology, Edward O. Wilson, was a particular target.[8]

Wilson's heresy was to have extended the theory of evolution to explain not just the shape but the behavior of organisms, which set off a new and productive strain of research that combined the genetic understanding of evolution with the study of animal behavior. Wilson secured the intense opposition of left-wing commentators when, following Darwin's example, he extended his theory to the behavior of human beings. The irony is that from the middle of the nineteenth until the late twentieth century evolution provided the scientific basis of progressive politics. Evolution is a theory of progress, after all, and the progress of life forms over the history of our planet was assumed to be reflective of the progress of Anglophone civilization and of human civilization in general; Marx had wanted to dedicate his great book on capital to Darwin (but Darwin refused the offer). By the end of the twentieth century, however, social evolution indicated certain irreversible strains in human nature such as gender differentiation, hierarchical social organization, and tendency toward violence. Once evolution had been extended to human behavior, progressivism had effectively lost its scientific support and was no longer intellectually viable. The implicitly conservative consequences of social biology were noted by popular British historian, Paul Johnson.[9]

The "wars" over sociobiology have since cooled down, and as the battles were well- reported at the time by "war correspondents" and lately have been reflectively recounted by historians, it is not necessary to itemize them any further. However, two other fronts in what has been a "war of the books" are worth discussing.

A mention needs to be made of the controversy surrounding *The Bell Curve: Intelligence and Class Structure in American Life* published in 1994, so overwhelming were the criticisms and attacked it provoked.[10] Its main thesis is that intelligence is a real, identifiable element of human personality, that it varies from person to person, that it is heritable to a definite and quantitative degree, and that it can be seen to vary not only among individuals but among gender and racial groupings. *The Bell Curve* then goes on to uncover the social effects of intellectual heritability, which it takes as sufficiently demonstrated. The point subsequently made in Herrnstein and Murray's book is its major point: that mathematical ability is becoming dominant in science and technology, which, in turn, is transforming society, and that such people who possess it are now living among themselves, (and more to the point, reproducing among themselves), with significant economic and social effects—chiefly, de facto segregation based on mathematical, ability.[11] This last point was lost in the furor that followed the book's publication, for what its authors took as well-proven its critics found most objectionable. And that is the assertion about intellectual ability as genetically transferable and its inevitable social impact.

The Bell Curve was attacked by progressives and left-leaning scientists, including Steven J. Gould, who had previously published his own book about heritable characteristics such as intelligence, which he claimed is not heritable. He went so far as to assert that "intelligence," in effect, was a false construction and did not actually exist as an empirically valid element of human nature.[12] *The Bell Curve* was attacked not as profoundly wrong on the sociological facts, but pernicious in its denial of one of the principle theses of modern progressivism, namely the essential equality of human beings, especially along gender and ethnic racial lines. The book was attacked as racist; it disturbingly ranged ethic groups according to their intelligence levels, as indicated by numerous types of intelligence test; Jews and Asians were first, followed by Caucasians, followed by blacks and Hispanics.[13] Critics of intelligence testing noted that IQ levels of various groups—Jews, for example—tested quite poorly when they first entered the United States as immigrants "off the boat," but that the same groups tested much higher in subsequent generations, demonstrating the inherent inadequacy of intelligence testing. (Thomas Sowell, a well-known conservative economist, made this point). The intent here is by no means to settle the contested issue of nature vs. nurture but to call attention to how, in this most elemental example of heritable intelligence, the thesis of a testable content within human nature so sharply conflicts with progressive ideals.

The last example of recent science to establish the permanence of certain unchangeable features of human nature is from the study of human cognition. In his book *The Blank Slate*, Steven Pinker, a neuropsychologist currently at Harvard (formerly at MIT), attempts to do two things at once. First, he explains the recent research, which shows that the human mind and resultantly human behavior has certain, highly defined traits that all human beings hold in common under all social environments. Secondly, he reassures progressives that the unchangeability of human nature does not subvert but in fact provides scientific support for the progressive agenda. It is fair to say that while Pinker succeeds in his first task, he does not do so in the second.[14]

Pinker begins by setting out what he says are three misunderstandings about the human mind—the "blank slate," the "noble savage," and the "ghost in the machine"—all of which are disqualified once the neuropsychological and evolutionary structures are understood. He states that the first two are the basis of left-wing political thought. Pinker denies that human personality begins with a *tabula rasa* on which society, learning, and experience imprint knowledge. He further denies that the effects of social influence on the mind are corruptive, in other words, that human nature is essentially good and noble when left unspoiled. These ideas are progressive, beginning at least with Rousseau. Pinker also denies the idea of the "ghost in the machine" to claim that the religious idea of an immortal soul implanted in the human body is another false idea about human nature. This false idea of the existence of the human soul Pinker sees as a basis of conservative political thought that he claims is understood to be the basis of morality.[15]

Pinker's book presents facts from across a wide front of contemporary biological and social science, which indicate that human nature is in large part a "given." Pinker is a discerning critic of the fallacious interpretations of human nature that underlie common beliefs, political ideals, and even scientific theories. He takes the presence of unchangeable structures as understood, and spends his efforts rather in presenting the relevant facts in such a manner as to argue that they do not present a challenge to progressive political thought. The main argument appears in Part 3, which consists of a series of chapters in which Pinker presents details about the topics of equality, perfectibility, determinism, and nihilism. He argues that progressive politics can not only survive this new knowledge about our species but thrive, once the points are well-enough understood.[16] Pinker follows up on the argument in Part 5, which deals with "hot button issues" such as gender and violence. In each case, he details how the new social science really does support a progressive ideology. But are Pinker's arguments, however well-expressed in a detailed, chummy, self-assured manner, really enough? They will not likely be reassuring to progressives once they realize that the scientifically established facts about human nature he presents, once accepted, sharply limit the possibilities of social change.

As for Pinker's portrayal of conservatism, the historical fact is that conservative doctrine has always rested on the notion that there is such a thing as human nature, which consists of certain permanent traits that invalidate the progressive ideal of society as the attainment of a utopian ideal. And, in fact, Pinker's portrayal of the constant elements in human nature, now discoverable by contemporary neurobiology and evolutionary psychology, tends to support the conservative understanding of human nature. While Pinker's understanding of the relationship between the doctrine of the soul and the origin of the human perception that there is a moral law is roughly on point, it ignores the fact that large number of progressives are religious believers, including mainline Protestants and liberal Roman Catholics. It is true as well that conservatives should be wary about too closely depending on recent evolutionary biology as a basis for their view of human nature, because it is reductive in a way that subverts the independent existence of moral rules, religious belief, and political ideals. Evolutionary psychology and neurobiology, if taken as the only true portrayals of the constancy of human nature, wield a sharp blade that cuts through the political ideals of both liberals and conservatives.[17]

Pinker states, in a hubristic manner that is all too characteristic of scientific writers, that the false notions of both progressivism and conservatism are now overcome by the "juggernaut" of science: "The belief on the left that human nature can be changed at will, and the belief on the right that morality rests on God's endowing us with an immaterial soul, are becoming rearguard struggles against the juggernaut of science."[18] It is a striking image, but not the first time that modern science has been portrayed in a triumphalist manner: Hobbes, Laplace, Spencer, Comte, Freud, Skinner, Wilson, Dawkins, and others have preceded Pinker. And yet a large cadre of the Unenlightened—the philosophers, novelists, poets, mystics, theologians, and humanistic psychologists—are still around. How odd it must seem that Whig, Enlightenment, and Reductive science cannot once and forever put away such false myths and mistaken philosophies, even among the well-educated! As for the figure of a juggernaut, it may be seen as a Panzer tank riding on mighty steel treads that leave deep ruts in the ground over which it runs on its course of inevitable triumph. Yet in the space between the ruts, the juggernaut leaves untouched the green grass of philosophic speculation, and the bright wild flowers of myth and religion. And among the growth in that tangled bank are the political ideals of progress, tradition, idealism, justice, freedom, equality and stability that are the bases of political philosophy and the social visions of the human future.

The Limits of Social Science—Nussbaum

Martha Nussbaum and Alasdair MacIntyre are two of the most prominent social philosophers writing at the present moment who have gone beyond the limits of social science to develop their philosophies; Nussbaum is seen as a "Liberal" philosopher because she has written on specific policy issues in an attempt to defend progressive positions, while MacIntyre, perhaps not with real accuracy, is seen as a conservative philosopher for his defense of traditions. For both philosophers, their primary resource has not been the research results of social science but instead a return to ancient sources that have emphasized the concept of virtue. The idea of human excellence prescinds from intellectualized definitions typical of modern ethical theory, including utilitarianism and deontology. Both philosophers, whose work represents an increasing trend in contemporary social philosophy, seek out manifestations in social reality of what is necessary and desirable in human behavior if a tolerable form of organized society is to exist; in short, the emphasis is on character, not the applicability of intellectual constructions.

MacIntyre denies to social science even the most basic qualification of being a scientific field, that of the ability of constructing valid general laws.[19] Nussbaum does not take up the issue of the validity or invalidity of the social sciences, but proceeds to develop her philosophy independently of the field. That is, while she relies on the results of psychological and sociological research to make a case regarding, for example, the place of women or minorities in society, her basic philosophy is based on a reading of ancient philosophers including, especially the Stoics and Aristotle. In her most important book, *The Fragility of Goodness*, Nussbaum rejects Plato's "science" in favor of Aristotle's approach. By "science" she means the construction of an integrated set of logical arguments in a deductive fashion, an approach that is comparable to the effort to discover a "'final' theory of everything." Such theories as the noble savage and economic determinism are the kind of deductive constructions that Nussbaum rejects. For Aristotle, ethical decision-making is not done in the manner of intuiting a major premise to which a specific kind of action is applied, and the conclusion as to its admissibility as a validly moral behavior is thereby determined. After a quote from Aristotle reflecting this point of view, Nussbaum comments:

> Aristotle argues here that the universal account *ought* to be regarded as only an outline, not the precise and final word. . . . It is not just that ethics has not yet attained the precision of the natural sciences; it should not even try for such precision. As applied to particular cases, which are the stuff of action, general scientific accounts and definitions are woefully lacking, of necessity, in the kind of suitedness to the occasion that good practice would require. [20]

But if these two highly regarded social philosophers philosophize from outside the rubrics of social science and empirical science generally, what are the problems within the social sciences that invalidate them as the primary source of social philosophy? There are three deficiencies: the variability of the data points, the tendency toward bias, and the descriptive as opposed to prescriptive aspect of social scientific laws. In the previous chapter, I put forth a law of diminishing explanatory returns based on the idea that as science proceeds up the ladder of ontological complexity, its results become progressively less secure. As a result of the variability of the data points, the social sciences are seen as the least secure. This is evident in the fact that generalizations within the social sciences are least often agreed upon by a majority of its practitioners, and that, as Kuhn was the first to make an issue of, the basic assumptions of what constitutes social science remain a constant subject of debate from within the social sciences themselves.[21]

It would be incorrect to state that the social sciences have an inevitable bias toward progressive social politics, but on the whole it is observably the case. The reason that many people enter fields including sociology and psychology is not only to study social phenomena in a scientific sense, but to change social and personal behavior. Thus, these fields become attractive to intelligent and active people who are also engaged in agenda-driven politics including racial justice, economic equality, and women's equality. In this frame of mind, of wanting not just to study social phenomena but to do good and improve society, it is inevitable that on a personal basis, interpretations of evidence, choice of research subjects, and modes of observation will reflect an ideological, that is, a progressive bias. And this has often been observed to be true.[22] Of course, some people who enter these fields are not ideological progressive; in fact, right-wing examples of social science can be found, for example, Charles Murray, Robert Nesbit, and George Homans, among sociologists.

The final limit on social science is the fact that its general laws and theories must, to the degree that they are properly scientific, function only as descriptions of patterns of human behavior and not as commands about how humans ought to behave. This indicates that social science is descriptive, not prescriptive. The foremost example of social science research is probably Myrdal's *An American Dilemma,* which is a comprehensive account of racial arrangements in the deep South of the United States in the 1940s, when Jim Crow laws were in effect. Myrdal began his study by pointing out the existence of certain American ideals, incontrovertibly demonstrating the conflict between the ideal of civil equality and the actual situation in America with regard to race. But Myrdal was aware that he could not command in a moral sense what Americans ought to do, for that is not the function of social science. Instead, he could only demonstrate the facts of the conflict between the ideal and the real.[23] In the social sciences, the Humean fix is that you cannot derive an *ought* from an *is*.

At a time when the progress of scientific technology seems an overwhelming aspect of our social lives—contemporary kitchens make ice, run oven fans, adjust room temperatures, and keep reminding the humans who cook in them when the roast is cooked—when, that is, technology seems to be taking on a life of its own, we need some standpoint from outside technological applicability to make decisions and evaluations on a human basis. But social science is too deeply implicated in the technologization of contemporary society to help much, for it has become the tool of manufacturers, "mad men" advertisers, and government officers.

The self-directed kitchen or house, however, is merely an annoyance; the situation is much worse when technology presents us with moral dilemmas. For it is guaranteed that as technologization proceeds, that *ethical dilemmas will increase in number and severity.* The moral dilemmas caused by technological development first became apparent to its current degree in World War II, when the finest research in physics since the eighteenth century in Relativity and Quantum mechanics brought about the reality of nuclear weapons. Currently, most examples are from biotechnology, as issues about cloning, infant stem-cell research, and human chimeras are now upon us, demanding moral clarification and standards. Even in electronic communication moral dilemmas arise, as in the *Aero* legal case that had to be resolved by a Supreme Court decision.[24] In that case, a device that took signals from cable television transmissions were sold very cheaply to bypass the fees (and profits) that should have gone to the producers of the entertainment. Despite claims that such a device was merely the legitimate development of new technologies, the Supreme Court decided, in effect, that it was not enough for technology to validate itself, and that theft is theft even when enabled by a new digital appliance. But the questions of where such ethical decisions come from and, especially, how they are to be validated cannot be answered by social science. Hence, the recent phenomenon has come into play of social philosophy going "back to the future," back to the portrayals of ancient virtue.

The Recovery of Ancient Virtue—MacIntyre

(Please note, Portions of this section originally appeared in *Modern Age*)

There is a nostalgic aspect to the philosophic analysis of progress that has been offered in these last two chapters, for even as it seeks to discover explanations of modern science in classical and modern philosophy, it is also destined to search for social values in the same venues. Contemporary social ethics has found a compelling source in the anciently defined doctrine of *virtue*. Virtue ethics is an expanding school of philosophy, as indicated by the recent appearance of scholarly books and articles. It is noteworthy that the recent turn to virtue ethics is not an explicitly conservative response in the political or ideological sense, but arises mostly because of the perceived inadequacies of those ethical theories that have appeared since Enlightenment times. As a theory, virtue ethics is based on the repudiation of the possibility of developing theories of ethics, including utilitarianism and deontology, as has been attempted since the Enlightenment. The historical nature of the case for virtue ethics originated from two formidable advocates, G.E.M. Anscombe and Alasdair MacIntyre. (Anscombe was one of Wittgenstein's disciples, the translator of his *Philosophical Investigations,* and a personal friend.)

Anscombe argued in favor of three propositions; first, "it is not profitable for us at present to do moral philosophy" and such efforts should be laid aside until at least a better understanding of human psychology is at hand; second, the concepts of *moral duty* and *moral obligation* and of "the moral sense of 'ought'" should be jettisoned; and third, the differences between well-known English writers on moral philosophy "are of little difference."[25] Anscombe's presentation is accessible only by intense study and the unpacking of her article, which is an implicitly historical account supplemented by complex analysis that assume the reader's ability to understand philosophical arguments in the mode of analytic philosophy. Nonetheless, the essence of the article is clear—that, since the end of the "dominance" of the Christian religion in Europe, ethical philosophy, has been and will continue to be a failure.

The thought of Alasdair MacIntyre on fundamental ethical issues has been expressed in a series of books that reflect the ethical, cultural, and political controversies of the contemporary age. As expressed in his later writings, MacIntyre's theory of ethics is based on the idea of *traditions*, which are the bearers or manifestations of the different and competing ethical ideas current in Western culture. It was developments in recent philosophy of science that had a direct impact on MacIntyre's idea of traditions as bearers of ethical thought. Kuhn's theory of *paradigms* and Lakatos' account of *research programmes* provided MacIntyre with a means to express clearly his basic idea that a particular ethical theory could not be properly understood without an intimate knowledge of the immediate social and historical contexts in which it was produced. MacIntyre's theory of ethical traditions in broad form translates *paradigms* and *research programmes* from the history of science as a means of understanding science, into the field of ethics. Of particular note is the idea of "untranslatability" in his later work that MacIntyre uses to describe the current situation in the West, where advocates of differing ethical and political traditions cannot find common ground for their debates. MacIntyre's thought here directly reflects Kuhn's well- known presentation of the "incommensurability" of scientific paradigms, a term that MacIntyre also uses.[26]

What Anscombe does implicitly, MacIntyre makes explicit: rejecting the universal ideal of reason as exemplified by Enlightenment thinkers, for example, Kant and Bacon. MacIntyre argues that any concept of universal reason in the Enlightenment sense is instantiated in a particular tradition or paradigm, since universal reason, if it exists, is not otherwise available to human beings and to the human mind, aside, that is, from a particular ethical tradition. MacIntyre argues that Enlightenment liberalism does not provide access to a universal reason by which to understand ethics and politics, for the ideal of universal reason is itself a tradition. Toward the end of Chapter VII of *Whose Justice*, MacIntyre makes the point: ". . . liberalism, which began as an appeal to alleged principles of shared rationality against what was felt to be the tyranny of traditions, has itself been transformed into a tradition whose continuities are partly defined by the interminability of the debate over such principles."[27]

After Virtue is MacIntyre's signal work. All his prior writings lead up to it, and his subsequent writings are developments of points made in it. The book has attracted wide attention and influence; its main intellectual contribution is to render the contemporary moral situation in the West comprehensible. As MacIntyre continually points out, citizens of the West today face an incoherent mix of ethical traditions that conflict and contend with one another. Many traditions contend, for example, those of "Kierkegaard, Kant, Diderot, Hume, [Adam] Smith," a situation that according to MacIntyre leads to profound moral confusion.[28] MacIntyre makes evident the depth of the moral confusion of our times by his initial presentation of the issue in *After Virtue*: ". . . my thesis entails that the language and appearances of morality persist even though the integral substance has to a large degree been fragmented and then in part destroyed."[29] His argument is based not only on a series of historical examples, but also on an interpretation of how theories of ethics develop in particular historical, social, political, cultural, and religious circumstances.

As a resolution to this moral confusion, which exists on both a theoretical and practical level, MacIntyre proposes that we go back to the ancient and precisely Aristotelian mode of doing ethical philosophy, and concentrate on the concept of *virtue*. Anscombe also makes this point, that rather than trying, in effect, to distill the essence of "justice" and to render an exact definition that can be applied generally, it is more useful to look at examples of what are without cavil seen as just actions by men who are universally acknowledged to be, because of a personal history of performing just acts, just men. In this view, more can be learned about justice by examining the careers of Lincoln and Washington than by a detailed comparison of Plato's *Republic* to Rawls' *Theory of Justice*. MacIntyre extends the method of historical analysis to develop the idea of ethical *traditions*, thereby describing each of the various theories of ethics in terms of their social and historical settings. MacIntyre's analysis concludes with a treatment of virtue as it is exemplified in each of these traditions, and then of both virtue and ethical traditions in themselves.[30] It is here that the complications in his theory of ethics arrive.

Since, in MacIntyre's view, an ethical tradition defines what virtue is according to the circumstances of a particular historical time and place, historical relativism becomes a major problem, for how does one decide which of the various contending traditions is the correct one? Dismissing Enlightenment rationalism as one of a variety of different traditions leaves no overall conception of ethical reality that is obviously available to apply as a universal standard. MacIntyre, however, is not a relativist, and, lacking a universally applicable criterion, he must find a means of judging comparative validity by looking within the traditions themselves. His solution is twofold. First, to declare that even though commitment to a tradition while living within it forecloses a commitment to other, rival traditions, advocates of a tradition can nevertheless be self-critical, acknowledging certain difficulties or contradictions within their tradition and recognizing the necessity for it to undergo further development if it is to succeed in future.[31] This self-critical aspect (which is not apparent in Kuhn's theory of paradigms) also makes it possible to acknowledge other traditions and to make comparisons among traditions.

The second part of MacIntyre's solution is to take his stance on the tradition that is Aristotelian, Thomistic, and Augustinian as the final choice among competing ethical traditions. MacIntyre is a Catholic convert and ended his career as Emeritus Professor at the University of Notre Dame. He justifies his choice by what could be called a criterion of *competitive inclusion*. He puts the question directly in a later book where he compares in great detail three major and contending ethical traditions, what he terms the "encyclopaedic," "genealogical," and "traditional" (a mixture of Aristotelian, Augustinian, and Thomistic influences). "Is there any way in which one of these rivals might prevail over the others? One possible answer was supplied by Dante: that narrative prevails over its rivals which is able to include its rivals within it, not only to retell their stories as episodes within its story, but to tell the story of the telling of their stories as such episodes."[32] Despite the complexity of his presentation, MacIntyre is arguing that his tradition is the truest one because it can explain and, in comparative terms, overcome any of the rival traditions by explaining them in their own terms as well as in the terms of his chosen tradition. Critics have pointed out that the dangers of postmodernist relativism still lurk in MacIntyre's comparative approach, and, further, according to Nussbaum, that his solution depends in the last analysis on religious faith. It may not be enough to argue the comparative intellectual superiority of a tradition that depends on a faith commitment, as MacIntyre does, yet his point is not a merely intellectual one; it appears these days that the futility of the different ethical approaches has played itself out in cultural terms. But he surely does not mean his immense contribution to ethical philosophy to be a disguised kind of religious prosletyzing, and given the general popularity of virtue ethics this is not thought to be the case.

Ethical theory has immediate application on a social level since it most often describes standards to judge how we relate to one another in a family or in society at large. Such questions as what parents owe to children or what the individual owes to the state, and vice versa, are all matters that are dealt with under the broad rubric of ethical theory. In the present circumstance, we have in the area of ethics a confusing mix of theories, including utilitarianism, moral relativism, egoism, and the like. Nussbaum's version of ancient Stoic virtue as the basis of a liberal state and a renewed cosmopolitanism, however, deserves special note since, like MacIntyre's theory of ethics, Nussbaum relies on a return to ancient virtue. In the American cultural mix, however, modern science also has a very large influence. But its dominance is failing, giving cultural space for the influence of new ideas that are in fact very old ideas, old and enduring.

Notes

1. http://plato.stanford.edu/entries/qm-manyworlds/
2. Barrow and Tipler, *The Anthropic Cosmological Principle*.
3. Stephen Hawking, *A Brief History of Time* (New York: Bantam, 1988), x.
4. Stephen Hawking and Leonard Mladunow, *The Grand Design* (New York, Bantam, 2010).
5. Lincoln Barnett, *The Universe and Dr. Einstein* (New York: Bantam, 1957), 109.
6. John Henry Newman, *A Grammar of Assent* (New York: Doubleday Image Books, 1955), 105.
7. Richard Dawkins, *the God Delusion* (Boston, Houghton Mifflin, 2006), 13, 114.
8. See E. O. Wilson, *Naturalist* (Washington, D.C., Shearwater Books, 1994), 336–53.
9. Paul Johnson, *Modern Times* (New York, Harper and Row, 1983), 733–34.

10. Richard J. Herrnstein and Charles Murray, *The Bell Curve* (New York, Free Press 1994).
11. Ibid., 51–61.
12. Stephen J. Gould, *The Mismeasure of Man* (New York, Norton, 1981). See Chapter Six, subtitled "Factor Analysis and the Reification of Intelligence, 234–320.
13. *Bell Curve*; in Chapter 13, "Ethnic Differences in Cognitive Ability", 268–315.
14. Steven Pinker, *The Blank Slate* (New York, Viking, 2002).
15. Ibid., 129–130.
16. Ibid., 137–280.
17. See John Caiazza, "Political Dilemmas of Social Biology" *Political Science Reviewer* Vol. XXXIV, 2005, 219–263.
18. Pinker, *Blank Slate*, 299.
19. Alasdair MacIntyre, *After Virtue,* 2nd ed. (Notre Dame: University of Notre Dame Press, 1984), 79–88.
20. Martha Nussbaum, *The Fragility of Goodness* (New York: Cambridge University Press, 2001, revised ed.), 303.
21. Kuhn, *Structure*, viii.
22. The Liberal bias in the social sciences recently observed by Jonathan Haidt. http://www.nytimes.com/2011/02/08/science/08tier.html
23. Gunnar Myrdal, *An American Dilemma* Vol. 2 (New Brunswick, Transaction, 1996).
24. *Aero* decision, http://techcrunch.com/2014/06/25/aereo-loses-in-supreme-court-deemed-illegal/.
25. G.E.M. Anscombe, "Modern Moral Philosophy," *Philosophy* 33(124): www. philosophy.uncc.edu/mieldrid/cmt/mmp.htn.
26. Alasdair MacIntyre, *Whose Justice, Which Rationality?* (Notre Dame: University of Notre Dame Press, 1988).
27. The point is also made in *Three Rival Versions of Moral Inquiry* (Notre Dame, U. of Notre Dame Press, 1990), in the criticism of encyclopedic rationality according to the traditional view.
28. *After Virtue*, 51.
29. Ibid., 5.
30. Ibid., 226–55.
31. WJWR, 349–369.
32. *Three Rival Versions*, 80, 81.

9

Conclusion: Crisis, Time, and the Choice

Crisis in Progress and Social Values

The arrival of a crisis in progress is more likely a harbinger of the old age of Western culture than it is of the lassitude of middle age, in which the uncounted reserves of youthful energy are no longer available. Western culture, having arisen in its modern form in the sixteenth and seventeenth centuries, has expanded its political, economic, and intellectual influence over the whole world by this time. It first remade itself from a more or less unified culture in the Middle Ages under the aegis of the church, into the diverse, competitive, individualistic, and scientifically materialist culture of modern times. However, in a person's middle age, when ambitions are (partially) fulfilled, the children grown, and retirement funds set aside, it is the time to pause and reflect. Where have we been, where are we going? In the cultural sense, we are now asking the same questions. As in life so in culture, as we understand that the cost of unalloyed ambition, even when the material events of history have been at our backs, has come we realize at some cost.

Our scientific ambitions first of all are under a new scrutiny. Chapters 2, 3, and 4 described versions of progress based on a historical view of science but with their inherent connections to varieties of philosophies of social progress. The presentation of the original Whig version, and the refinements of the Enlightenment and Reductive variations, was an attempt to explain progress in all its variety and social consequences. Chapter 5, however, described a more recent take that denies the reality of scientific progress by describing its theoretical output not in terms of propositions acquired by means of close observation and organized experiment, but as whole points of view, world-views, research *programmes*, and paradigms that envision a scientific field as a whole entity. And in this version, change in scientific theory quite accurately resembles the acquisition of a political ideology or a religious conversion. But it also requires that the expression of scientific commitments is as much psychological and social as a matter of deduction. The epistemological consequence is that truth ultimately is relative. This development is the initial sign of the crisis of progress as we have known it since the Enlightenment, and its social consequence is a kind of relativism of social values as well. It is at this point that the crisis is seen in its full theoretical mode. The complementary aspect is that technology, as argued in Chapter 6, is at this same point in time producing not progress but stasis in our culture as the agent of fulfillment of middle-class lifestyles.

The idea that contemporary technology is a harbinger not of continued progress but of cultural stasis seems perhaps contrarian, but evidence that the culture has in fact come to a pause is evident from the general manner in which technology is currently applied. Any application of any new technology today is immediately considered in cost/benefit terms, which was not often been done in the past; industrialization and technology have literally changed the face and atmosphere of the planet. The best example of cost/benefit considerations regarding new technologies today is fracking, the extraction of gas and oil from shale and rock rather than in deep deposits underground. Fracking, because it takes place on the surface, has direct effects on the environment immediately surrounding the fracking sites. Therefore, while on one hand it is touted as an unalloyed good since it may enable the United States to become independent of foreign sources for its oil and gas, on the other hand environmental concerns have led the federal government to prohibit the technology from federal lands. In the past, the unconsidered applications of technologies have had negative, if unanticipated, effects. A current example of this is the effect of overreliance on antibiotic medicines to fight infections, which has led—in proper Darwinian fashion—to the development of "superbugs" that cannot be eliminated by traditional medicines like penicillin.

Chapters 7 and 8 attempt to revise the understanding of progress by looking it from a philosophical rather than a scientific point of view. That, in turn, leads to an appreciation that the recent arrival of physical cosmology as a legitimate field of scientific inquiry has philosophic consequences that cannot be contained, as it were, in an empiricist vessel. Much of recent popular writing by scientific advocates however seems to want to make the point, of containing the obvious nonempiricist consequences of recent cosmological speculation. In this way, only materialistic science has the controlling word, the final say, on the implications of the current wave of scientific speculation on how many universes there are, whether our universe had a beginning, and whether human life and life itself was somehow intended as a consequence of evolution. Why should scientists have all the fun?

The major consequence of the pause in progress is that thinking about social values has transcended the limits of empiricist scientific explanation in favor of philosophic and literary approaches. This can be seen most clearly in the return to virtue. This was a concern of ancient philosophy and literature (Aristotle and Homer), of the Middle Ages (Aquinas and Dante), and has now become a concern in our own time for both theoretical and practical reasons (Nussbaum and Tolkien).

Agentic Time

It is generally understood that science in the twentieth century has radically altered our view of space, but the change in our concept of time has been less well-understood. The concept of time has changed a great deal from when Newton described it as flowing "equably" in the eighteenth century. Thermodynamics arose during the nineteenth century because of the practical necessity of improving steam engines, which were being developed for mining applications and transportation. Among the consequences of heat theory is entropic decay, which is observable in practice because of the impossibility of retaining steam power once the energy of overheated vapor had been dissipated by powering pistons. The inevitable decay of steam energy is described mathematically as the consequence of a fully developed theory, The Second Law of Thermodynamics. Thus, time itself was conceived not as a neutral plane on which Newton's laws of motion played out, but rather has a direction (entropy was called "time's arrow"), which when applied to the universe as a whole indicates the inevitable state of final decay.

The evolutionary concept of time lends itself to a more optimistic understanding, for the long passage of geological time is seen as a platform of steady improvement and of the arrival of new life forms. The span of time itself had to be extended so that the history of life on earth could be measured not in thousands, or millions, but in billions of years. Fossil evidence indicated not only that an enormous period of time had passed but also that there had been a progression in the levels of complexity of life forms—from single-celled organisms, to primitive life forms such as trilobites, to ancient fish species, to the long-dead giant dinosaurs, to prehuman Neanderthals. Evolutionary time was cosmic as well in that it was connected first to the geology of the planet earth, but then backward to the formation of the sun and solar system, then the Milky Way galaxy, and finally the ultimate origin in time of the universe itself. From the big bang theory until about 100,000 earth-years ago, the evolutionary account is enormously long, progressive, and time-bound.

But in considering the evolution of life, the question of its first origins on earth presents evolutionary theory with a complex and thorny problem. For natural selection and variation cannot as a matter of principle explain the origin of life itself, since the appearance of the first form of life on earth from the time that the earth itself comes into existence is less than 500 million years. This is an eye blink in evolutionary terms and does not provide enough time for the long-term processes that should presumably have taken place for life itself to have first appeared. This problem seems to afflict the explanation of evolution overall, for the sudden appearance of new forms of life in the fossil record indicates not slow processes but rapid ones, more quantum effects than evolutionary ones.[1]

I have argued that time is not a neutral plane on which biological processes take place and new forms of life slowly evolve. In order to make up for the otherwise inexplicable facts of sudden origins, time itself is necessarily a causative factor, hence an agent.[2] And what is true of biological evolution may be applied to the evolution of the universe at large., Its time sequence is a causative element in its development, from a single infinitely charged particle to hot gases, to the formation of galaxies (second-generation stars in which the heavier elements were formed), to the formation of planetary systems, and finally, and not incidentally, to our own earth. Retrospectively, cosmic evolution appears as a time-bound process in which the passage of time produces sudden and dramatic effects. It is not a neutral plane on which all the important activities take place as if on a stage set; the universe and all things in it are time-bound.

Given that time should be understood not as a mere sequence on a neutral plane, we can ask whether on the historical level *the present affects the past* and in this way realign events into a perceivable pattern. If we admit the thesis, it would help explain certain empirically validated events such as prophecies. And if we limit the thesis to admit only that simultaneity replaces time-sequenced events, then the operation of physical cause-and-effect does not require a span of measurable time and we are in the area of Bell's theorem and the EPR experiment where time disappears. If so, we as human beings are involved ourselves in the creation of time, and the creation of histories becomes re-creation. Here we are on the speculative level of popular science fiction perhaps, but more relevant to the purpose of this book is the inference that human beings have a responsibility to write histories, not as reimaginings but as re-creations.

Thus, we are in a time when the Enlightenment belief in progress must be reconsidered and its history reevaluated, rewritten, and re-created. It is not a matter of disowning the concept of Enlightenment progress based on the advance of science and progressive social ideals, but of rewriting them, re-creating them in order to become fully aware of the nature of the crisis before us. This point is an extension into the twenty-first century of Kant's *Ninth point* in chapter 3, that is the obligation to "write a general world history" vis a vis the Enlightenment.[3]

The Nature of the Crisis: Pascal or Nietzsche?

The crisis of progress, the theme of this book, once again needs a redefinition. What we face or are experiencing is a form of existential crisis. It is a soft form of crisis, for it is based on the existential fact that the culture, not only Western culture but the newly developing world culture as well, is facing a decision. We are at a decision point that can be expressed as the choice between Nietzsche or Pascal, both known as existential writers. Thus, it is important to understand that the options are not between religious value and scientific materialism because we live in a postmodern age that relies on meta-analysis and so the options must be more subtly defined. Such a dilemma between science and religion would be easy, too easy in fact, and would miss the point. Our materialism is based on the sumptuous levels of technology that are aimed at making our middle-class lives even easier than they are now; technology is devoted far less to quantum improvement in human life than to gradual easing. The goal is not so much to make a fulfilled life possible but to provide greater convenience for the middle class. Our experience of material comfort renders the option of materialism somewhat less objectionable than if it were offered as a philosophical option, as in scientific materialism as expressed by Dawkins, Weinberg, or even the ancient atomists. On the other hand, the religious option is in fact hard to pin down for it offers a large variety of different options, ranging from a compelling but vague sense that there must be "something beyond" the material universe described by the avatars of modern science, to Christian fundamentalism, which requires a strict interpretation of all the clauses of the King James Bible. And, to follow one of the major themes of this book, each of these options has a direct influence, is entangled with, a particular set of social values. What we face, as elaborated by Pascal and Nietzsche, respectively, is the choice between an unsettling search for transcendence or the agitated expressions of self-empowerment.

The Pascalian option starts with Pascal's understanding of the science of his time, an understanding that has not gone out of date despite the vast progress that has been made in science since the seventeenth century. Pascal famously wrote that the silence of the infinite spaces of Newton's universe frankly terrified him, devoid as they were of any human component whatsoever—indifference brought to an extreme. The scientific universe since that time has become more complex, mainly because we now understand it as composed of undergoing transitions of development and change through time. But, if it makes any sense, the universe has become even more infinite than it was in Pascal's time, though it is no longer silent. Rather, the universe in post-Newtonian cosmology has a voice, it seems, or several competing voices. One says that humankind did not arrive on the evolutionary scene by accident, but that our arrival was somehow intended. The other voice, the one that Pascal heard—in effect, the "voice of silence"—says that the human species is really of no more account in the evolutionary sense than are whales or redwood trees.

But whether in Newton's universe or in our own, Pascal's primary concern remains the same and is pertinent: What is the place of mankind and, reflecting further, what is my place in the vast universal frame of things? How can I find purpose there, and a place for myself as I live my life in its daily rounds and seek long-term meaning or clarity? What has modern science to say to me, and to mankind generally, about these philosophical and in the personal sense existential matters? Some scientific advocates write as if, in considering the universe as modern science presents it, we ought to be awestruck, and indeed we are, but what if, as in Pascal's case (not an unusual one), that awe turns to terror? What if the universe is Lovecraft's version, in which the indifference of the universe is personified as an alien entity that seems to hate us? What if, as in Nietzsche's statement, we look into the abyss and the abyss looks back? What then? Here begins what is familiarly termed "the search for meaning," which may mean flinging ourselves at any remote possibility that implies that the universe has a care for us. That, in a way, is the motive for Pascal's bet, as if he were saying, "What have you got to lose? You may as well act as if God exists, even if in some small, unrepentant portion of your intellect you still retain a doubt."

But the rise of cosmology gives rise to the calculations that vote against the probability of the evolution of the human race; there were so many essential but unconnected steps that had to fall into place in just the exact manner. These calculations also vote against the cosmos as it is understood by contemporary science, the hard sciences—that is, physics, chemistry, and cellular biology—that must provide within the advanced equations and intricate operations of nature itself a place that allows for human existence. Einstein's contribution was as much a matter of the imagination as of the rational intellect, for now we can envision internally consistent alternate universes that are just as worthy of existence in the mathematical realm as our own. But in that case, the inevitable question arises as to why *this* universe, the homey one we humans inhabit exists, and not another one that is too hot or too cold or that does not contain carbon rings, solid planets, or the other possible myriad characteristics that would prohibit the existence of life in general, much less the advanced form of human life? In that question, put forward by the anthropic principle that is as much a technical as a philosophic question, lays the hope for the discovery of meaning that Pascal could not (yet) see. Nowadays, Pascal's bet need not be only a matter of will, a frantic search for meaning that rests on a chance, but a rational option as well.

The crisis of progress is a matter of options, however, and, like Pascal's, Nietzsche's option has a cosmological aspect if not a motive. For Nietzsche's science was that of the ancient atomists in which the history of the universe is a great circle, calculated by the ancient Hindus as 10,000 years (in one account), the concept of eternal recurrence. The point of the great circle route to cosmology is that in effect the universe is eternal, has always existed from eternity and always will. Furthermore, the actions of individual human beings and the history of civilizations are repeated also, and such a cosmic repetition has the effect of denying such events their sense of moral imperative or value. Actions and history repeated in this cosmic manner have no permanent consequence, and thus are denuded of their moral worth; all human action, whether individual or social, is only part of a great cosmic wheel, the snake eating its own tail, as in Aztec mythology. The myth of eternal return is echoed in a precise manner by the speculations of current cosmologists about a recurring universe, the point of which is that the material universe is eternal, never ending, and without a discoverable beginning; it simply always was, is, and will be. Such a cosmos obviously has divine pretentions.

Yet the point of Nietzsche's speculations about eternal recurrence are not really motivated by an attempt to develop a cosmology; rather, eternal recurrence is a means of denuding actions that are traditionally judged by religious or philosophic standards as morally nugatory and worthless.

This, in turn, leaves open only one motive for purposive human action: the putting forth of oneself as, in effect, a cosmic principle. Here we have the nut of Nietzsche's purpose, which allows us to trace its relevance to our civilization's present state of stasis. For what enables the current civilization's emphasis on self—self-development, personal health, the "bucket list," a successful love life and career, and self-expression to its maximum degree—is not Nietzsche's philosophy of the *Ubermensch*, but rather technology. As explained in chapter six, technology today promises much less often new horizons and technological jump-starts than a constant development of conveniences and minor technological achievements that support the middle-class lifestyle. The release of yet another new model of the cell phone is not a technical development that will revolutionize our world; that was done 100 years ago by Alexander Graham Bell and Samuel F. B. Morse, whose inventions made possible instantaneous information-sharing that was independent of geographical barriers. Subsequent developments have mostly been a matter of increased convenience.

Nonetheless, current technology does enable a person to become, as it were, a super-man or superwoman on their own, for nowadays everyone can be their own *Ubermench*. Contrary to Nietzsche's original vision, the technological version of the *Ubermensch* is egalitarian; no one "super-person" is superior to any other. This is an odd conclusion that is made possible by the retail technologies that are intended to give the consumer the impression that only he or she matters, and by the technologies that allow us to live independently from society at large, without the need for personal interaction. It is peculiar that Nietzsche's nihilistic vision of the transvaluation of values, including the destruction of national destiny, religious faith and philosophic rationality, is now translated in American middle-class culture as a source not of existential fear but of personal comfort. Surely something is being lost here, but we somehow seem unable to identify it; perhaps as Nietzsche implied, we should be afraid to look into the abyss.

So, which option do we choose, Nietzsche or Pascal? This question is not only personal but cultural. Despite its provenance, which implies the ultimate rejection of cultural norms, choosing the Nietzschean option is actually a choice for stasis. This option depends on technology and advances in science to enable us to keep in place, like running assiduously on a treadmill. The amount of effort and the feeling of reaching previously unexplored heights of experience and self-knowledge is only the runner's high that kicks in when body chemistry is slightly altered.

However, the Pascalian option also has its deficiencies, for it places the person and the culture in a place that is not static but that remains open to the possibility of a transcendent experience, which necessarily upsets the culture whenever it is expressed. Instead of the abyss, we may discover and be confronted by alien presences that are benign and yet as disconcerting as the discovery of the abyss, for they might make demands on us to participate in their own life.

Notes

1. See Francis Crick, *Life Itself: Its Origin and Nature* (New York: Simon and Schuster, 1981).
2. John Caiazza, *The Ethics of Cosmology* (New Brunswick, NJ: Transaction, 2012).
3. For a recent example of a reconsideration, see Vincenzo Ferrone trans. E. Tarantino, *The Enlightenment: History of an Idea* (Princeton, 2015).

Index